Physics of the Body

$(.3927)(222.22)$

$(87.266)/(2.026 \times 10^{-13})$

(87.266)

$A = \pi r^2$

$r^2 = \dfrac{A}{\pi}$

$r = \sqrt{\dfrac{A}{\pi}}$

Physics of the Body

John R. Cameron
University of Wisconsin–Madison

James G. Skofronick
Florida State University–Tallahassee

Roderick M. Grant
Denison University, Granville, Ohio

Medical Physics Publishing
Madison, Wisconsin

20 19 18 17 16 15 14 13 12 11   6 7 8 9 10 11 12

The second edition of this book has been translated into Greek (ISBN 960-394-102-6) in 2001 and into Korean (ISBN 89-86865-54-8) in 2002.

The first edition of this book is a revision of about half the material in *Medical Physics* by John R. Cameron and James G. Skofronick, a Wiley-Interscience Publication. © 1978 by John Wiley & Sons, Inc.

Chapter 6 of this book is based on *Intermediate Physics for Medicine and Biology*, 3rd Edition, by Russell K. Hobbie. © 1997 by Springer-Verlag New York, Inc., and is used with permission.

Library of Congress Cataloging-in-Publication Data

Cameron, J.R. (John Roderick), 1922–
    Physics of the Body / John R. Cameron, James G. Skofronick,
Roderick M. Grant. — 2nd ed.
       p.   cm.
    Rev. ed. of: Medical physics / John R. Cameron, James G.
Skofronick, Roderick M. Grant. c1992.
    Includes bibliographical references and index.
    ISBN 0-944838-90-1 (hardcover). — ISBN 0-944838-91-X (softcover)
    1. Medical physics.   I. Skofronick, James G.   II. Grant, Roderick
M.   III. Cameron, J.R. (John Roderick), 1922–   Medical physics.
IV. Title.
    [DNLM: 1. Biophysics.   2. Physiology.   QT 34 C1817m 1999]
R895.C28   1999
612'.014—dc21
DNLM/DLC
for Library of Congress   99-26114
                                         CIP

ISBN 0-944838-90-1
ISBN 0-944838-91-X (softcover)
ISBN-13: 978-0-944838-91-4

Cover illustration created by Lauren Shavell and Pat Carrico, Medical Imagery.

Medical Physics Publishing
4513 Vernon Boulevard
Madison, WI 53705-4964

Printed in the United States of America

This book is dedicated to our readers—
we hope they enjoy learning about how their bodies work.

Contents

3   Muscle and Forces   37

7 Physics of the Lungs and Breathing   143

8 Physics of the Cardiovascular System   181

9 Electrical Signals from the Body   215

10 Sound and Speech 257

11 Physics of the Ear and Hearing 279

12 Physics of the Eyes and Vision   307

Preface

Physics was called Natural Philosophy in the 17th and 18th centuries. The word "physics" comes from the Greek word for nature and forms the root word of a number of medical words such as physician, physical, and physiology. One of the first physics texts to include physics of the body was *Elements of Physics or Natural Philosophy: General and Medical* by Neil Arnott, M.D., published in London in 1827. It marked the beginning of medical physics as a coherent discipline. The book had a profound impact on the teaching of physics in the 19th century and on the development of medical physics in the 20th century. It was very popular and went through six editions. It was used as a physics text for about 75 years.

The scientific field of physics covers matter, forces, and energy in various forms. Physics has two important areas of application in medicine—the physics of the body (the physics of physiology) and the physics of instruments used for diagnosis and therapy. In this book we discuss the physics of the body. We will occasionally describe a medical instrument that is related to the organ being described.

An antique name for the physics of the body is iatrophysics. Iatro is from the Greek word for medical. The first iatrophysicist was the Italian physicist, Giovanni Alfonso Borelli. His book *De motu animalum* (On the motions of animals) was published posthumously in 1680. It treats the movements and contractions of muscles and includes his explanation of the power of a torpedo fish to produce shocks by the rapid contractions of a muscle now known as the electric organ. Later in the 17th and early 18th century, Galileo and his medical friend Sanctorious applied physics to the development of some simple medical instruments—a thermometer to indicate the temperature of the body and a simple pendulum to measure the rate of the pulse.

While the roles of chemistry and biology in medicine are well accepted, the role of physics is usually not as obvious. Although most medical and paramedical students take an introductory physics course, they often see little or no applications of physics to medicine. This communication gap is due primarily to the large amount of material in the traditional college physics course that precludes an adequate treatment of physics of the body. A second reason is that, in general, physics teachers have very little understanding of the physics of the body.

This book is written primarily for students who plan to make a career in some field of medicine. We describe in a simple fashion the usefulness of physics in understanding the function of the various organ systems of the body such as the eyes, ears, lungs, and heart. Although this book was written primarily as a text for students who have some understanding of elementary physics, we believe much of it will be interesting and understandable to any person who is curious about how his or her body works. The mathematics is at the algebra level.

This new edition of *Physics of the Body* is an update of the first edition, which in turn was a revision of about half of the material in the book *Medical Physics* by Cameron and Skofronick, published by Wiley-Interscience in 1978. (*Medical Physics* is out of print and the copyright was returned to the authors.) This new edition incorporates improvements in several chapters, including the addition of several problems. The order of the chapters has been modified, new material on forces in dentistry has been added, and the physics of the ear and hearing is now a separate chapter with more detail on the function of the inner ear.

The International System (SI) of units is used throughout the book, but the more common units are usually given in parentheses. This book is not intended to be a reference book, but a selection of general references as well as a bibliography specific to each chapter topic has been added at the end of the text. Appendix A gives data on the Standard

Man. Logarithms and exponentials are used in a few places; they are reviewed in Appendix B.

Emeritus Professor Russell Hobbie of the University of Minnesota, the author of the more advanced text *Intermediate Physics for Medicine and Biology*, kindly contributed Chapter 6 on the physics of osmosis as it relates to fluid transport across membranes in the body. He also contributed to the revision of Chapter 9. His cooperation is greatly appreciated.

We again thank our friends who helped with the predecessor book *Medical Physics* who are acknowledged in the preface to that book. We also thank Ms. Joy Opheim for her contribution to the dentistry portion concerning the physics in teeth. We give special thanks to Betsey Phelps, Editor of Medical Physics Publishing, for her efforts to get this book published.

We thank our wives Von Cameron, Dot Skofronick, and Sue Grant for their patience and understanding while we worked on the book.

John R. Cameron
James G. Skofronick
Roderick M. Grant

Gainesville, Florida
Tallahassee, Florida
Granville, Ohio
May 1999

1

Terminology, Modeling, and Measurement

In this chapter we present four broad topics. First we examine what we mean by medical physics and describe some related and overlapping disciplines. In Section 1.2 we discuss modeling, a concept that is essential in science, engineering, and medicine. In Section 1.3 we discuss and give examples of feedback, an important feature of many models. Finally, in Section 1.4 we briefly discuss precision and accuracy in measurements using examples drawn from studies of the human body.

1.1 Terminology

The term *medical physics* refers to two major areas: (1) the applications of physics principles to the understanding of the function of the human body in health and disease and (2) the applications of physics in the instrumentation used in diagnosis and therapy. The first of these could be called the physics of physiology; the second is often referred to as clinical medical physics.

1

The word *physical* appears in a number of medical contexts. The words *physicist* and *physician* have a common root in the Greek word *physike* (science of nature). Only a generation ago in England a professor of physic was actually a professor of medicine. Today the first thing a physician does after taking a medical history of a patient is a physical examination. During this examination, the physician uses the stethoscope to listen to heart, lung, and blood sounds in the arteries; measures the pulse rate and blood pressure; and in other ways applies principles and techniques derived from physics. The branch of medicine referred to as *physical medicine* deals with the diagnosis and treatment of disease and injury by means of physical agents such as manipulation, massage, exercise, heat, and water. *Physical therapy* is the treatment of disease or bodily weakness by physical means such as massage and exercise rather than by medications or surgery.

In principle, the field of *biophysics* should include medical physics as an important subspecialty. However, biophysics has evolved into a relatively well-defined field that has very little to do with medicine directly. It is primarily involved with the physics of large biomolecules, viruses, and so forth, although it does approach medical physics in the study of the transport of materials across cell membranes. Biophysicists conduct basic research that may improve the practice of medicine in the next generation. Most medical physicists engage in applied physics in the diagnosis and treatment of disease. A few medical physicists do research to improve the practice of medicine in the current generation.

The field of medical physics is concerned with the concepts and methods of physics as applied to the treatment of human disease. Some medical physicists work as teachers and scientists in universities and medical schools, or in industry. However, most are directly involved in patient care in hospitals in a field called *radiological physics*. That is, they are involved with the applications of physics to radiological problems, such as the use of ionizing radiation (x-rays, γ-rays, electrons, and/or radionuclides) in the diagnosis and/or treatment of disease. Others work on problems related to the application of diagnostic procedures using tools from such diverse areas of applied physics as magnetic resonance imaging (MRI), ultrasound (e.g., echo cardiography or sonography), and lasers. Physicists identifying themselves with any of the activities listed above usually have doctorates of philosophy (Ph.D.s) with advanced training specific to their roles. Often they are identified professionally through membership in the American Association of Physicists in Medicine (AAPM). The AAPM has more than 4,000 members.

1.2 Modeling

Even though physicists believe that the physical world obeys the laws of physics, they are also aware that the mathematical descriptions of some physical situations are too complex to permit solutions. For example, if you tore a small corner off this page and let it fall to the floor, it would perform various gyrations. Its path would be determined by the laws of physics, but it would be impossible to write the equation describing this path. Physicists would agree that the force of gravity would cause it to go in the general direction of the floor if some other force did not interfere. Air currents and static electricity both could affect its path. Similarly, although the laws of physics are involved in all aspects of body function, each situation is so complex that it is impossible to predict exact behavior using our knowledge of physics. Nevertheless, a knowledge of the laws of physics will help our understanding of physiology in health and disease.

Sometimes in trying to understand a physical phenomenon we simplify it by selecting its main features and ignore those that we believe are less important. Our description may be only partially correct, but it is better than none at all. In trying to understand the physical aspects of the body, we often resort to analogies. Keep in mind that analogies are never perfect. For example, in many ways the eye is analogous to a TV camera; however, the analogy is poor when one considers 3D vision.

In this book we often use analogies to help us explain some aspect of the physics of the body. We hope that we are successful, but please remember that all explanations are incomplete to some degree. The real situation is always more complex than the one we describe. Many of the analogies used by physicists employ models—model making is common in scientific activities. A famous nineteenth-century physicist, Lord Kelvin, said, "I never satisfy myself until I can make a mechanical model of a thing. If I can make a mechanical model I can understand it!"

Some models involve physical phenomena that appear to be completely unrelated to the subject being studied; for example, a model in which the flow of blood is represented by the flow of electricity is often used in the study of the body's circulatory system. This electrical model can simulate very well many phenomena of the cardiovascular system. Of course, if you do not understand electrical phenomena, the model does not help much. Also, as mentioned before, all analogies have their limitations. Blood is made up of red blood cells and plasma. The percentage of the blood occupied by the *red blood cells* (the hematocrit) changes as the blood flows toward the extremities. This phenomenon (discussed in Chapter 8) is difficult to simulate with the electrical model.

Other models are mathematical; equations are mathematical models that can be used to describe and predict the physical behavior of some systems. In the everyday world of physics we have many such equations. Some are of such general use that they are referred to as laws. For example, the relationship between force (F), mass (m), and acceleration (a)—usually written as F = ma—is known as Newton's second law. There are other mathematical expressions of this law that may look quite different to a layperson but are recognized by a physicist as other ways of saying the same thing. Newton's second law is used in Chapter 2 in the form F = Δ(mv)/Δt, where v is the velocity, t is the time, and Δ (delta) indicates a small change of the quantity. The quantity mv is the momentum, and Δ(mv)/Δt is to be read as rate of change of momentum with respect to time. This form of Newton's second law is identical to F = ma.

1.3 Feedback

One of the physicist's favorite words is *function*. The symbol for function, *f*, should not be confused with the symbol for force F. The equation W = f(H) means the weight W is a function of the height H. It does not tell you how weight and height are related or what other factors are involved. It is sort of a mathematical shorthand. In the medical field we could write R = f(P) to indicate that the heart rate R is a function of the power produced by the body P. The next step—to leave out the *f* and write an equation that describes specifically how the things are related to each other—is the hard one.

A medical researcher may use a model of some function of the body to predict properties that were not originally suspected. On the other hand, some models are so crude that they are only useful for serving as guides to improved models.

Many parameters of the body remain relatively constant (e.g., stability of temperature or calcium in the blood). *Homeostasis* is the medical word that describes the tendency for the body to maintain this stability. In physics or engineering this is called *feedback control*. An engineer who wants to control a quantity that changes with time will take a sample of what is being produced and use this sample as a signal to control the production to some desired level. That is, some of the output is fed back to the source to regulate the production. If the system is designed so that an increase in the output results in a signal feeding back which decreases the production and that a decrease in the output results in a signal feeding back which increases the production, the feedback is negative. Negative feedback produces a stable control, while positive feedback, in which a change in the sample fed back causes a change in the same direction, produces an unstable control.

An example of an important negative feedback mechanism (homeostasis) in the body is control of blood glucose (blood sugar). A normal human maintains the level of glucose at about 850 mg/liter. (This is usually given as 85 mg/dl). That is, about 0.085% of your blood is glucose or in a typical adult there are about five grams of sugar—about one-sixth ounce. Glucose is the fuel for the brain. Like oxygen, it is necessary for life and is continually being used up. Too high or too low a glucose level causes serious health problems. When you eat, your blood glucose goes up within about 30 minutes. The pancreas senses the higher level and releases insulin into the blood to speed up the consumption of glucose by the cells to maintain it near the proper level. The pancreas of a type 1 diabetic (a condition sometimes called "juvenile diabetes") is unable to produce insulin and the level of glucose may become many times higher than normal. This can cause serious health problems. The body attempts to regulate the glucose level by other means. The most obvious way to reduce glucose, if there is no insulin, is for the kidneys to "dump" some of the excess sugar into the urine along with more water. This causes the urine to be sweet and the loss of water leads to frequent urination and thirst. The serious long-term risks of excess glucose are damage to the circulation that sometimes leads to amputations of feet, blindness, defective kidney function, and nerve damage. Until 1923, when insulin was discovered, life expectancy of type 1 diabetics was very short.

When a normal healthy person isn't eating, the glucose is reduced by normal metabolism. It can fall below its normal level. A sensor in the pancreas triggers the feedback control to excrete another hormone, glucagon, to stimulate the liver to release glucose into the circulation. This feedback control usually keeps the blood glucose at about 800 mg/liter. This type of glucose production can go on for many days if one does not eat. The pancreas of a type 1 diabetic does not produce glucagon and the low blood sugar—hypoglycemia—quickly leads to a confused mental state in less than an hour. This is one of the first signs of hypoglycemia. It also leads to loss of muscular control, which can be a great risk if you are driving at the time. If severe hypoglycemia continues, it will lead to coma and eventually death if the glucose level is not increased. Some diabetics have been mistaken for being drunk when they lose muscular control.

The normal healthy person has feedback mechanisms to keep the glucose level from going too high or too low. A type 1 diabetic, however, cannot control either problem. By injecting insulin before eating diabetics are able to crudely control high glucose levels. They can control low blood glucose levels by eating small portions of food more often than a person without diabetes would eat. In an emergency, if the diabetic cannot take food, it is possible to inject glucagon to temporarily stimulate release of glucose by the liver. It is better for the diabetic to take food before the level gets too low. There have been big improvements in the control of diabetes, such as the invention of the individual glucose tester. This tester requires a drop of blood from the diabetic, over the course of the day, to regulate blood sugar. The glucose tester is, however, a poor substitute for the sophisticated feedback sensor and control mechanisms of the pancreas.

Negative feedback control is common in the body. Another important function of the body is to control the level of calcium in the blood. If the level drops too low, the body releases some calcium from the bones to increase the level in the blood. If too much calcium is released, the body lowers the level in the blood by removing some via the kidneys. Although many of the control mechanisms of the body are not yet understood, various diseases have been found to be directly related to the failure of these mechanisms. For example, as the body grows, its cells keep increasing in number until the body reaches adult size, and then it remains more or less constant in size under some type of feedback control. Occasionally some cells do not respond to this feedback control and become tumors.

1.4 Measurement

One of the main characteristics of science is its ability to reproduce measurements of quantities of interest. In the practice of medicine, early efforts to measure quantities of clinical interest were often scorned as detracting from the skill of the physician. For example, even though body temperature and pulse rate could be measured during the seventeenth century, these measurements were not routinely made until the nineteenth century. During the past century there has been a steady growth of science in medicine as the number and accuracy of quantitative measurements used in clinical practice have increased. It can even be said that many measurements are now made more accurately than is necessary for clinical purposes. Often it is the change of a quantity over time that is important. That is, precision (repeatability) is more important than accuracy. (See the discussion on precision and accuracy later in this section.)

Figure 1.1 illustrates a few of the common measurements used in the practice of medicine. Some of these measurements are more reproducible than others. For example, an x-ray gives only qualitative information about the inside of the body; a repeat x-ray taken with a different machine may look quite different to an untrained observer.

In an introductory physics course many different types of measurement are studied. In general, International System (SI) units, or metric units, are used to measure various quantities. Unfortunately, they are not in common use in the United States as they are in most of the world. The basic SI units we use in this book are the meter (m) for length, the kilogram (kg) for mass, and the second (s) for time. For convenience we often use larger or smaller units. For clarity, we may use both the

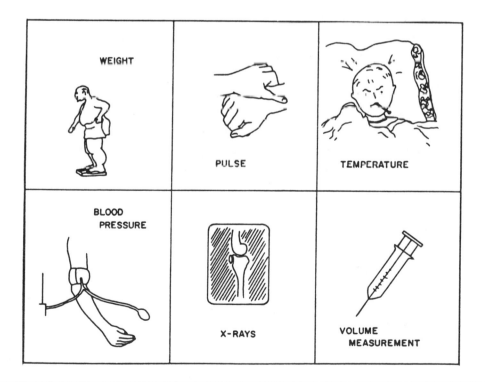

Figure 1.1. Some common measurements in the medical field.

larger and smaller units for better understanding. Some examples are given below:

$$\text{centimeter (cm)} = 10^{-2} \qquad \text{``one-hundredth of a meter''}$$

$$\text{millimeter (mm)} = 10^{-3} \qquad \text{``one-thousandth of a meter''}$$

$$\text{micrometer (}\mu\text{m)} = 10^{-6} \qquad \text{``one-millionth of a meter''}$$

$$\text{nanometer (nm)} = 10^{-9} \qquad \text{``one-billionth of a meter''}$$

$$\text{gram (g)} = 10^{-3}\,\text{kg} \qquad \text{``one-thousandth of a kilogram''}$$

$$\text{milligram (mg)} = 10^{-3}\,\text{g} \qquad \text{``one-thousandth of a gram''}$$
$$\text{or } 10^{-6}\,\text{kg} \qquad\qquad \text{``one-millionth of a kilogram''}$$

We sometimes give quantities in English units such as feet, pounds, and gallons for easy visualization, and we sometimes use the units common to a particular branch of medicine in the United States.

The use of a unit implies that there is a "true" or standard unit to which measurements with that unit can be compared. Most modern countries have laboratories that specialize in the standardization of measuring systems. In the United States, the National Institutes of Standards and Technology (NIST) located in Gaithersburg, MD, performs this function. NIST has little contact with medicine, but it has played an important role in standardizing the measurement of ionizing radiation in the treatment of cancer.

In medicine in the United States, many quantities are expressed in nonstandard units. For example, while the SI unit for pressure is pascal (Pa) or force per unit area (in newtons per square meter), blood pressure in the United States is generally expressed in millimeters of mercury (mm Hg), a height of a column of liquid! This length is the height of a column of mercury that has a pressure at its base equal to the blood pressure. We discuss pressure further in Chapters 5, 7, and 8.

A nonstandard medical unit that is no longer used but is of historical interest is length to represent pulse rate. In 1585 while a medical student, Galileo discovered the principle of the pendulum while watching a chandelier swing in church. He used his pulse to time the period of the swing and found that the time for one large swing was the same as the time for one small swing. This led to his development of the pendulum clock. Shortly thereafter (in 1602) Sanctorius, a medical friend of Galileo, invented the pulsologium (a simple pendulum) to measure the pulse rate of his patients. An assistant adjusted the length of the pendulum until it swung in time to the pulse rate as called out by Sanctorius. The pulse rate was then recorded as the length of the pendulum.

PROBLEM 1.1

Show that your resting pulse occurs at a time interval that corresponds to the period of a pendulum whose length is somewhere in the range L = 0.1 to 0.25 meters. [The period of a simple pendulum is given by $T = 2\pi(L/g)^{1/2}$, where L is the length of the pendulum and g is the acceleration due to gravity.]

There are many other physiological measurements involving time. We can divide them into two groups: measurements of repetitive processes, such as the pulse, and measurements of nonrepetitive processes, such as how long it takes the kidneys to remove a foreign substance from the blood. Nonrepetitive time processes in the body range from the duration of the action potential of a nerve cell (~1 ms) to the average life span of an individual (80 yr = 2.5×10^9 s or 2.5 Gs where G = giga for 10^9 or 1 billion).

PROBLEM 1.2

Determine your age in minutes. Include an estimate of the error in your answer.

Measurements of repetitive processes usually involve the number of repetitions per second, minute, hour, and so forth. For example, an average pulse rate is about 70 min^{-1} and an average breathing rate is about 15 min^{-1}. Different repetition rates in the electrical signals from the brain, caused by mental activity, range from less than 1 s^{-1} to more than 20 s^{-1}.

Many subtle processes in the human body are repetitive. For example, the body has a circadian rhythm (approximately 24 hours) that is present even when a person is living in a deep cave without any way of telling the time. Another biologically controlled clock is the very slow running clock that governs a growing person's mass. This clock causes the rate of growth to fluctuate with a period of 14 to 18 months.

In science *accuracy* and *precision* have different meanings. Accuracy refers to how close a given measurement is to an accepted standard. For example, a person's height measured as 1.765 m may be accurate to 0.003 m (3 mm). However, that same person's height may vary over the course of the day by as much as 0.005 m (5 mm), so there is no accepted standard even for one individual. Precision refers to the reproducibility of a measurement and is not necessarily related to the accuracy of the measurement. For example, a patient's temperature was measured ten times in a row with the following results (on the Celsius scale): 36.1, 36.0, 36.1, 36.2, 36.4, 36.0, 36.3, 36.3, 36.4, and 36.2. The accepted standard, normal body temperature, is 37 C or 98.6 °F. The precision of this measurement was fairly good, with a range of measured values of ±0.2 C from the average value of 36.2 C. However, when compared with a recently calibrated standard thermometer, the thermometer was found to be defective, reading 3 C low. This inaccurate thermometer would not be satisfactory for clinical use; even with a fever, the patient's temperature would be reported as normal.

1.3

PROBLEM

Verify that the *average* variation of the range of measured temperatures in the preceding paragraph is ±0.1 C.

In general, it is desirable to have both good accuracy and good precision. However, sometimes an accurate measurement cannot be obtained even with a measuring technique that has good precision, and in such cases the precise but inaccurate measurement may be useful. For example, the defective thermometer discussed previously could be used to determine if a patient's temperature was stable, rising, or falling. Sometimes the accuracy is limited by uncontrollable factors; for example, it is difficult to measure internal parts of the body accurately, such as the amount of mineral in bones. One technique for measuring the bone mineral has a precision of approximately 1% but an accuracy of 3 to 4% (see Chapter 4). The technique is nonetheless very useful for following changes in the amount of bone mineral such as occur in treatment for osteoporosis.

It is an accepted fact in science that the process of measurement may significantly alter the quantity being measured. This is especially true in medicine. For example, measuring the blood pressure may introduce errors (uncertainties); one factor, called the M.D. effect, may be related to the anxiety of a patient in the doctor's office. This type of error may also be introduced in taking a patient's history. The patient may not truthfully answer questions dealing with personal matters, such as sexual practices or alcohol or drug consumption. Using computers for patient interviews may help to reduce this type of error.

A clinical measurement by itself does not necessarily determine whether a patient is well or ill. For each medical test there is usually a well-established range of normal values. In addition, the values just above and below the normal range are usually considered equivocal; no decision can be made with certainty from these values. Finally, the results of many clinical laboratory tests can be affected by outside factors such as medications.

After a physician has reviewed a patient's medical history, the findings of the physical examination, and the results of clinical laboratory measurements, a decision must be made as to whether the patient is ill and, if so, what the illness is. It is not surprising that sometimes wrong decisions are made. These wrong decisions are of two types: false positives and false negatives. A false positive error occurs when a patient is diagnosed as having a particular disease that is in fact not present; a false negative error occurs when a patient is diagnosed as free of a particular disease when in fact the patient has the disease.

In some situations a diagnostic error can have a great impact on a patient's life. For example, a young person who was thought to have a rheumatic heart condition spent several years in complete bed rest before it was discovered that a false positive diagnosis had been made—in fact, the patient had arthritis, a disease in which activity should be maintained to avoid joint stiffening. In the early stages of many types of cancer it is easy to make a false negative diagnostic error because the tumor is small and difficult to detect. Since the probability of cure depends on early detection of the cancer, a false negative diagnosis can greatly reduce the patient's chance of survival.

Diagnostic errors (false positives and false negatives) can be reduced by researching the causes of misleading laboratory test values and by developing new clinical tests and better instrumentation. Errors or uncertainties from measurements can be reduced by using care in taking the measurement, repeating the measurements, using reliable instruments, properly calibrating the instruments, and by making care-givers aware of human error.

To illustrate how measurement errors can be reduced, let us consider the problem of determining a person's weight accurately and precisely. (You might want to perform this experiment yourself in order to become more familiar with this type of analysis.)

Our first willing subject was asked to stand on a bathroom scale while we carefully read the scale with a magnifying glass. Ideally the individual should have on minimal clothing so that we would not have to weigh the clothes separately in order to correct the measurement for their weight. The subject was asked to step off and on the scale several times to see if the same weight was measured each time. We found small variations that were largely the result of different positions of the feet on the scale. The results using the bathroom scale are given in column 2 of Table 1.1. The average (mean) weight was 156.3 lb, with two-thirds of all the measurements between 156.0 and 156.6 lb. This result can be written as (156.3 ± 0.3) lb. The ±0.3 lb is often referred to as the uncertainty or the *standard deviation* (SD) of the measurements. (To learn more about standard deviations and related matters, see, e.g., Chapter 13 of the book by Stibitz listed in the bibliography.)

It occurred to us that the bathroom scale might not be very accurate, as it was quite old. As a comparison we decided to also use two of the new scales at the local school gymnasium and one from the campus hospital for the experiment. The results obtained on these scales are given in columns 3, 4, and 5 of Table 1.1. The means and uncertainties for the results at the gym and the hospital are shown to be, in pounds, (155.2 ± 0.2), (155.5 ± 0.2), and (156.0 ± 0.1). The hospital scale is clearly more precise, as the results indicate. However, it seems best not to draw any conclusions about the relative accuracy of any of the sets of scales since we did not know our subject's activities (eating, exercise, defecation, etc.) between the various data gathering sessions.

An observation at the gym bears consideration. We noticed that the indicator of our subject's weight jiggled even though every attempt was made to stand still. We hypothesized that the jiggling might be due to the beating of our subject's heart. Each time the heart beats, it forces a mass of blood upward, which forces the body downward (Newton's third law), resulting in an apparent momentary increase in weight with each heartbeat. As the bulk of the rushing blood reaches the first bend in the

Table 1.1. Results from Several Weighings on Different Scales

Weighing	Bathroom Scale	Gym Scale#1	Gym Scale#2	Hospital Scale
1	156.3	155.0	155.6	156.0
2	155.9	155.2	155.4	156.2
3	156.0	155.3	155.3	155.9
4	156.7	155.4	155.5	155.8
5	156.8	155.1	155.7	156.1
6	156.1	155.5	155.2	156.0
7	156.3	154.9	155.8	155.9
8	156.2	155.3	155.4	156.1
9	156.4	155.1	155.7	—
Mean	156.3	155.2	155.5	156.0
SD	0.3	0.2	0.2	0.1

aorta and heads toward the feet, it has the opposite effect and causes an apparent small decrease in weight. We found in fact that our subject's pulse was synchronized with the jiggles. Try it yourself!

Calibration of scales to make them more accurate is relatively unimportant—an error of even one pound would not make much difference. However, this research aroused our interest in other sources of error. Someone asked about the effect of breathing on body weight. Does a person weigh more when the lungs are full or when they are empty? We decided to try the experiment. Our subject weighed the same both after a deep breath and after a forced expiration, but we noticed that the weight appeared to jiggle during deep breathing. Air weighs about 1 g/liter, and our lungs hold about 5 liters. Since there are 454 g in 1 lb, it follows that the weight of the air in the lungs would be about 0.01 lb, much less than we could hope to see with our uncertainty of 0.1 lb.

Then someone else remembered Archimedes' principle. Since the body is in a "sea" of air, it is buoyed up by the weight of the air displaced by the body. Thus it does not make any difference whether the air is inside or outside the lungs. We decided to calculate the buoyant effect for a 70 kg (154 lb) person. The body is about as dense as fresh water (that is why you can just float in fresh water), and 1 kg of water has a volume of 1 liter (10^{-3} m^3). Thus a 70 kg person has a volume of about 70 liters. Since a liter of air has a mass of about 1 g, the buoyant effect on the body would be about 70 g or roughly 0.15 lb—again, this would not be seen.

This exercise still did not explain why the weight of our subject appeared to change during breathing. Then we recalled that when we breathe in, the diaphragm lowers, causing the center of gravity to lower slightly; when we breathe out, it rises slightly. The acceleration of the center of gravity downward at the beginning of an inspiration causes a momentary apparent reduction in weight; the acceleration of the center of gravity upward during expiration causes a similar increase (Fig.1.2). This explains the effect, and indeed we found that a larger effect was produced when our subject breathed in and out more rapidly.

In summary: (1) all measurements contain uncertainties and are somewhat inaccurate, (2) with special effort we can reduce the error and the uncertainty, and (3) in many cases there is no need to improve the measurement because the quantity being measured is naturally variable.

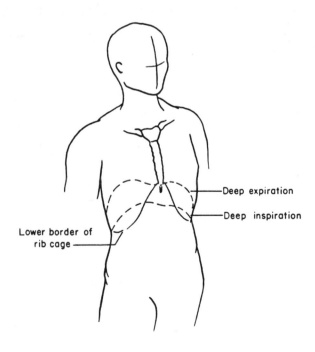

Figure 1.2. Levels of diaphragm during deep inspiration and deep expiration. Upon inspiration, the center of gravity is lowered slightly.

1.4

PROBLEM

(a) Describe a method you might use for measuring the height of a person who is still growing. What are the sources of error?
(b) Is measuring height the best way of determining growth changes? Can you think of other methods?

1.5

PROBLEM

Measure your pulse 10 times for 15-second periods. (a) What is your average pulse rate per minute? (b) Estimate the accuracy and the precision of this measurement. How can the accuracy be improved?

1.6

PROBLEM

What is the ratio of your pulse rate to your breathing rate? (It is usually about 4:1.)

2

Energy, Heat, Work, and Power of the Body

Energy is a basic concept of physics. In the physics of the body energy is of primary importance. All activities of the body, including thinking, involve energy consumption. The conversion of energy into work such as lifting a weight or riding a bicycle represents only a small fraction of the total energy use of the body. Under resting (basal) conditions the body's energy consumption is used principally by: the skeletal muscles and the heart (25%); the brain (19%); the kidneys (10%); and the liver and spleen (27%). The remaining 19% is distributed over many systems, such as digestive.

Food is the body's basic source of energy (fuel). The food we consume is generally not in a form suitable for direct energy conversion. It must be chemically changed by the body to make molecules that can combine with oxygen in the body's cells. We do not discuss this complex chemical process—the Krebs cycle. From a physics viewpoint we can consider the body to be an energy converter that is subject to the law of conservation of energy. The body uses energy from food to operate its various organs, to supply heat to maintain a constant body temperature, to do external work, and to build a stored energy supply (in the form of fat) for later needs. A small percentage (~ 5%) of the food energy is excreted in the feces and urine. The energy used to operate the organs eventually appears as body heat. Some of this heat is useful in maintaining the body at its normal temperature, but the rest must be disposed of. Other energy sources, such as heat from the sun and heat energy from our surroundings, can help maintain body temperature, but are of no use in body function.

In this chapter we discuss the conservation of energy in the body (first law of thermodynamics), the conversion of energy in the body, the work done by and the power of the body, and how the body controls its temperature and loses heat.

2.1 Conservation of Energy in the Body

Conservation of energy in the body can be written as a simple equation:

$$\begin{bmatrix} \text{change in stored} \\ \text{energy in the body (i.e.,)} \\ \text{food, energy, body fat,} \\ \text{and body heat)} \end{bmatrix} = \begin{bmatrix} \text{heat lost} \\ \text{from the body} \end{bmatrix} + \begin{bmatrix} \text{work done} \end{bmatrix}$$

This equation, which is a statement of the *first law of thermodynamics*, assumes that no food or drink is taken in and no feces or urine is excreted during the interval of time considered.

There are continuous energy changes in the body both when it is doing work and when it is not. We can write the first law of thermodynamics as

$$\Delta U = \Delta Q - \Delta W \tag{2.1}$$

where ΔU is the change in stored energy, ΔQ is the heat lost or gained, and ΔW is the work done by the body. (By Convention, ΔQ is positive if heat is added *to* the body, and ΔW is positive if work is done *by* the body.) A body doing no work ($\Delta W = 0$) and at a constant temperature in general will lose heat to its surroundings if the surroundings are at a lower temperature, and thus ΔQ is negative. Therefore, ΔU is also negative, indicating a decrease in stored energy. The energy term, ΔU, is discussed in Section 2.2, the work term, ΔW, is discussed in Section 2.3, and the heat term, ΔQ, is considered in Section 2.4.

It is useful to consider the rates of change of ΔU, ΔQ, and ΔW (the change in these quantities in a short interval of time Δt). Equation 2.1 then becomes

$$\frac{\Delta U}{\Delta t} = \frac{\Delta Q}{\Delta t} - \frac{\Delta W}{\Delta t} \qquad (2.2)$$

where $\Delta U/\Delta t$ is the rate of change of stored energy, $\Delta Q/\Delta t$ is the rate of heat loss or gain, and $\Delta W/\Delta t$ is the rate of doing work; that is, the mechanical power. Note that all three terms necessarily have units of energy/time, or power.

Equation 2.2, which is used extensively in this chapter, is another form of the first law of thermodynamics. It tells us that the rate of energy change is conserved in all processes, but it does not tell us whether or not a process can occur. For example, the first law tells us that if we put heat into the body at a certain rate, $\Delta Q/\Delta t$, we could expect the body to produce or store chemical energy at that rate, or to produce a certain rate of work output. This does not occur; the physical law describing the direction of the energy conversion process is known as the *second law of thermodynamics*. The second law also limits the fraction of stored energy, ΔU, which can be converted to useful work, ΔW.

2.2 Energy Changes in the Body

The SI unit for energy is the newton-meter (Nm) or joule (J); power is given in joules per second (J/s) or watts (W). Because physiologists and nutritionists use kilocalories (kcal) for food energy and kcal/min for the rate of heat production, we retain these units in this book. It should be noted that the kcal is the same as the food Calorie you read about on packaged food products and in articles about nutrition. A diet intake of 2400 Calories/day is thus the same as 2400 kcal/day. Since 1 cal = 4.184 J, 2400 kcal/day ~1×10^7 J/day. There are 86,400 s/day, so the average power P~115W.

The relationships between some of these units are summarized as follows:

1 kcal = 4184 J (= 1 Calorie)

1 kcal/min = 69.7 W

100 W = 1.43 kcal/min

1 kcal/hr = 1.162 W

Lavoisier was the first to suggest (in 1784) that food is oxidized after consumption. He based his arguments on measurements of an experimental animal that showed that oxygen consumption increased during the process of digestion. He explained this effect as work of digestion. We now know that this explanation is incorrect; the correct explanation is that oxidation occurs in the cells of the body.

In oxidation by combustion, heat is released. In the oxidation process within the body, heat is released as energy of metabolism. The rate of energy production is called the *metabolic rate*.

Let us consider the oxidation of glucose—$C_6H_{12}O_6$—a common form of sugar used for intravenous feeding, and the principal source of energy for the brain. The oxidation equation for glucose, in moles, is:

$$C_6H_{12}O_6 + 6O_2 \rightarrow 6H_2O + 6CO_2 + 2.87 \times 10^6 \text{ J} \qquad (2.3)$$

That is, 1 mole of glucose (0.18 kg) combines with 6 moles of O_2 (0.192 kg) to produce 6 moles each of H_2O (0.108 kg) and CO_2 (0.264 kg), releasing 2.87×10^6 J of heat energy in the reaction. Using this information we can compute a number of useful quantities for glucose metabolism. (A *mole* is the amount of a substance that contains as many atoms or molecules, ions, or other elementary units as the number of atoms in 0.012 kilogram of carbon 12. (Remember that 1 mole of a gas at normal temperature and pressure has a volume V = 22.4 $\times$ 10^{-3} m^3 = 22.4 liters.)

Energy released for each kg of glucose =
$2.87 \times 10^6/(0.180) \cong 16$ MJ/kg
(where M = "mega" = 10^6 or one million)

Energy released for each m^3 of O_2 used =
$2.87 \times 10^6/(6 \times 22.4 \times 10^{-3}) = 21$ MJ/m^3

Volume of O_2 used per kg of glucose =
$6 \times 22.4 \times 10^{-3}/(0.180) = 0.75$ m^3/kg

Volume of CO_2 produced per kg of glucose =
$6 \times 22.4 \times 10^{-3}/(0.180) = 0.75$ m^3/kg

Similar calculations can be done for fats, proteins, and other carbo-hydrates. Typical caloric values of these food types and of common fuels are given in Table 2.1. Table 2.1 also lists the energy released per liter of oxygen consumed for the various types of food. By measuring the oxygen consumed by the body, we can get a good estimate of the energy produced.

Note that the total mass on the two sides of the equation 2.3 is the same (0.372 kg). However, since 2.87 MJ of heat was released, the total mass on the right hand side is slightly less by an amount determined by Einstein's relation between mass and energy, $\Delta E = \Delta mc^2$.

$$\Delta m = \Delta E/c^2 = 2.87 \times 10^6 \text{ J}/(3 \times 10^8 \text{ m/s}^2)^2 \sim 3 \times 10^{-11} \text{ kg}$$
$$\text{(compare to 0.372 kg)} \qquad (2.4)$$

In Table 2.1 the values given for the foods are the maximum that might be expected. Not all of this energy is available to the body because part is lost in incomplete combustion. The "unburned" products are released in feces, urine, and flatus (intestinal gas). What remains is

Table 2.1. Typical Energy Relationship for Some Food and Fuels

Food or Fuel	Energy released per unit volume of O_2 consumed (J/m^3)	Energy released per kilogram consumed (J/kg)	Energy released per gram (kcal/g)
Glucose	21.0×10^6	1.6×10^7	3.8
Carbohydrates	22.2×10^6	1.72×10^7	4.1
Proteins	18.0×10^6	1.72×10^7	4.1
Fats	19.7×10^6	3.89×10^7	9.3
Typical Diet	$20.1–20.9 \times 10^6$	—	—
Gasoline	—	4.77×10^7	11.4
Coal	—	3.35×10^7	8.0
Wood (pine)	—	1.88×10^7	4.5

energy that can be metabolized. The body is usually quite efficient at extracting energy from food. For example, the energy remaining in normal feces is only about 5% of the total energy contained in the consumed food. Since the body is at constant temperature, the energy in the consumed food plus the body fat make up the available energy.

When completely at rest, the typical person consumes energy at a rate of about 92 kcal/hr, or about 100 W. This rate of energy consumption, called the *basal metabolic rate* (BMR), is the amount of energy needed to perform minimal body functions (such as breathing and pumping the blood through the arteries) under resting conditions. Clinically an individual's BMR is compared to normal values for a person of the same sex, age, height, and weight. The BMR depends primarily upon thyroid function. A person with an overactive thyroid (*hyperthyroid*) has a higher BMR than a person with normal thyroid function.

The BMR test was replaced by the 24-hr thyroid uptake test. The thyroid uses iodine in the hormone that controls metabolic function. The amount of iodine taken from the diet is a good measure of metabolic activity. It is easily measured with radioactive iodine. After World War II radioactive iodine (I-131) with an 8-day half-life became available. It is possible to measure the "uptake of iodine" as an indicator of thyroid function. A small dose (37 kBq) of I-131 is given to the patient by mouth and the next day the amount of radioactivity in the thyroid is measured with a scintillation detector. The percent uptake is determined by comparing the radioactivity in the neck with that from an identical I-131 capsule in a plastic cylinder the size of the neck.

Since the energy used for basal metabolism becomes heat which is primarily dissipated from the skin, one might guess that the basal rate is related to the surface area or to the mass of the body. Figure 2.1 is a plot of BMR (expressed in kcal/day) for various animals of widely different weights. The slope of the line indicates that the BMR is proportional to $(mass)^{3/4}$. Thus as animals get larger, their BMR increases faster than their surface area, which is proportional to $(mass)^{2/3}$, but not as fast as their volume (mass). The metabolic rate depends to a large extent on the temperature of the body. Chemical

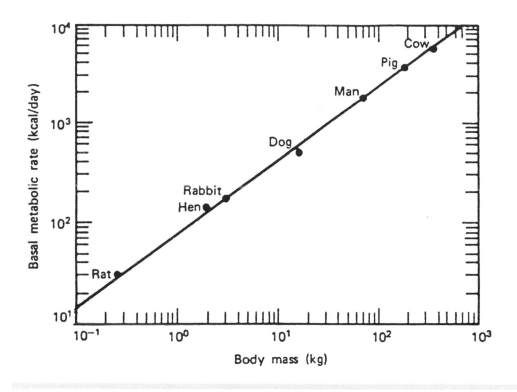

Figure 2.1. Relationship between the basal metabolic rate (BMR) and the body mass for several different animals.

processes are very temperature dependent—a small change in temperature can produce a large change in the rate of chemical reactions. If the body temperature changes by 1 C, there is a change of about 10% in the metabolic rate. For example, if a patient has a temperature of 40 C, or 3 C above normal, the metabolic rate is about 30% greater than normal. Similarly, if the body temperature drops 3 C below normal, the metabolic rate (and oxygen consumption) decreases by about 30%. You can see why hibernating at a low body temperature is advantageous to an animal and why a patient's temperature is sometimes lowered during heart surgery to reduce oxygen consumption.

Obviously, in order to keep a constant weight an individual must consume just enough food to provide for basal metabolism plus physical activities. Eating too little results in weight loss; continued too long, eating too little results in starvation. A diet in excess of body energy needs will cause an increase in body fat (weight).

The BMR used to be determined from the oxygen consumption when resting. We can also estimate the food energy used in various physical activities by measuring oxygen consumption. Table 2.2 gives some

Table 2.2. Typical Oxygen Consumption and Power Needed for Everyday Activities

Activity	O$_2$ Consumption × 10^{-6} (m^3/s)	Equivalent Heat Production kcal/min	Equivalent Heat Production J/s (W)	Energy Consumption (J/m^2 s)
Sleeping	4.0	1.2	83	47.7
Sitting at rest	5.7	1.7	120	66.8
Standing relaxed	6.0	1.8	125	72.6
Riding in a car	6.7	2.0	140	78.5
Sitting at a lecture (awake)	10.0	3.0	210	119.1
Walking slowly (5 km/hr)	12.7	3.8	265	151.1
Cycling at 15 km/hr	19.0	5.7	400	226.6
Playing tennis	21.0	6.3	440	250.0
Swimming breaststroke (1.6 km/hr)	22.7	6.8	475	265.0
Skating at 15 km/hr	26.0	7.8	545	310.0
Climbing stairs at 116 steps/min	32.7	9.8	685	390.0
Cycling at 21 km/hr	33.3	10.0	700	395.0
Playing basketball	38.0	11.4	800	450.0
Harvard Step Test*	53.7	16.1	1120	640.0

*A test in which the subject steps up and down a 0.4 m step 30 times/min for 5 min.

typical values for various activities. Here, the rate of energy consumption is given in J/(m^2 s); this makes allowance for people of different sizes. To find your body surface area in m^2, use the empirical relationship: A = 0.202 M$^{0.425}$H$^{0.725}$ where H is your height in meters and M is your mass in kilograms (ref: Ruch and Patton).

Oxygen consumption for various organs has been measured, and these values are given in Table 2.3. Note that some of the organs use rather large amounts of power and that the kidney uses more power per kilogram than the heart.

Table 2.3. Oxygen Use and Metabolic Rate Contribution by Principal Organs of a Resting, Healthy 65 kg Man*

Organ	Mass (kg)	Avg Rate O$_2$ Consumption by Experiment (ml/min)	Power Consumed (kcal/min)	Power per kg (kcal/min)/kg	Contribution as % of BMR
Liver and Spleen	—	67	0.33	—	27
Brain	1.40	47	0.23	0.16	19
Skeletal muscle	28.00	45	0.22	7.7×10^{-3}	18
Kidney	0.30	26	0.13	0.42	10
Heart	0.32	17	0.08	0.26	7
Remainder	—	48	0.23	—	19
		250	1.22		100

*Adapted from R. Passmore, in R. Passmore and J. S. Robson (Eds.), *A Companion to Medical Studies*, Vol. I., Blackwell, Osney Mead, England (1968).

Energy consumption, including weight loss resulting from dieting or physical exercise, is explored in the following problems:

PROBLEM 2.1

Suppose you wish to lose 4.5 kg (10 lb) either through physical activity or by dieting.

(a) How many hours would you have to work at an activity of 10^3 J/s (~1 kW) to lose 4.5 kg of fat? (Of course, you could not maintain this activity rate very long.) (Hint: see Table 2.1.)
[Answer: T = 2810 min ~ 47 hr.]

(Note that a great deal of exercise is needed to lose a few pounds. Of course, you could not continue this activity level for very long, let alone 47 hr. To verify this claim, compute the power you need to run up a flight of stairs a vertical distance of 3 m in a time of 3 s. Assume your mass is 50 kg.)
[Answer: P = ~0.5 × 10^3 J/s = 0.5 kW]

(b) It is usually much easier to lose weight by reducing your food intake. If your normal diet is 2400 kcal/day, how long must you diet at half that value (1200 kcal/day) to lose 4.5 kg of fat?
[Answer: T = ~35 days.]

2.2 **PROBLEM**

For a hypothetical animal that has a mass of 700 kg,
(a) Use Fig. 2.1 to estimate the basal metabolic rate of this animal.
   [Answer: ~ 10^4 kcal/day]
(b) Assuming the animal's food provides 5 kcal/g, estimate the mass of food needed by this animal each day.
   [Answer: 2.0 kg]

2.3 **PROBLEM**

By what percent does your metabolic rate increase if you have a fever 2C above normal?

2.4 **PROBLEM**

(a) What is the energy required to walk a distance of 50 km at a rate of 5 km/hr. [Answer: 9.5×10^6 J or 2270 kcal.]
(b) Assuming an energy equivalent of your food of 2.1×10^7 J/kg (5 kcal/g), calculate the amount of food needed for the walk. [Answer: 0.45 kg ~1 lb].

2.3 Work and Power

Chemical energy stored in the body is used to support life-preserving functions and is converted into external work. It is clear that external work is done when a person is climbing a hill or walking up stairs. In this case, we can calculate the work done by multiplying the person's weight in N (W = mg) by the vertical distance (h) moved in m. When a person is walking or running at a constant speed on a level surface, most of the forces act in the direction perpendicular to the motion. It appears that the external work done is zero (since W = Fx, where F is the component of force parallel to the displacement, x). However, muscles in the leg are doing internal work which appears as heat in the muscle and causes a rise in its temperature. This additional heat in the muscle is removed by blood flowing through the muscle, by conduction to the skin, and by sweating. These processes are considered in Section 2.4.

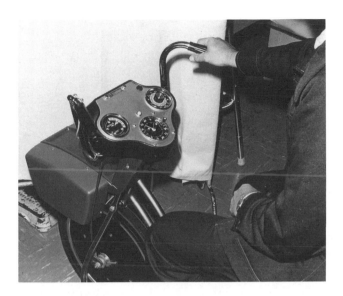

Figure 2.2. The ergometer, a stationary bicycle with adjustable friction that permits studies of oxygen consumption under various work loads. One of the meters indicates the power produced.

In this section we discuss the efficiency of the human body as a machine for doing external work. This topic lends itself well to experiment. For example, we can measure the external work done and power supplied by a subject riding on an ergometer, a fixed bicycle that can be adjusted to vary the amount of resistance to the turning of the pedals. We can also measure the oxygen consumed during this activity. The total food energy consumed can be calculated since ~20 kJ are produced for each m^3 of oxygen consumed.

The efficiency ϵ of the human body as a machine is:

$$\epsilon = \frac{work\ done}{energy\ consumed}$$

Efficiency is usually lowest at low power but can increase to 20% for trained individuals in activities such as cycling and rowing. Table 2.4 gives mechanical efficiencies for various activities and for several engines.

Studies have shown that cycling is one of our most efficient activities (see Problem 2.5). For a trained cyclist the efficiency approaches 20% with an external power production of 370 W and a metabolic rate of 1850 W. If the cyclist is on level ground and moving at a constant speed, there is no change in potential or kinetic energy and the power supplied is used primarily to overcome wind resistance and friction of tire flexing.

Table 2.4. Some Typical Mechanical Efficiencies

Task or Machine	Efficiency (%)
Cycling	~20
Swimming on surface	<2
Swimming underwater	~4
Shoveling	~3
Steam Engine	17
Gasoline Engine	38

The maximum work capacity of the body is variable. For short periods of time the body can perform at high power levels, but for long-term efforts it is more limited. Experimentally it has been found that long-term power is proportional to the maximum rate of oxygen consumption in the working muscles. For a healthy man at rest oxygen consumption is typically 4×10^{-6} m^3/kg of body weight each minute. At maximum activity it is typically 4×10^{-5} m^3/kg (4 ml/kg) of body weight each minute

The body can provide energy for short-term power needs by splitting energy-rich phosphates and glycogen, leaving an oxygen deficit in the body. This process can only last about a minute and is called the anaerobic (without oxygen) phase of work; long-term activity requires oxygen (aerobic work). Figure 2.3 shows these phases of work for a cyclist.

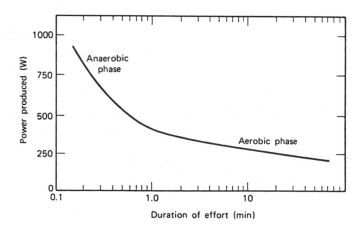

Figure 2.3. Typical power output on a bicycle for a healthy adult.

PROBLEM 2.5 Compare the energy required to travel 20 km on a bicycle to that needed to travel by auto for the same trip. Gasoline has an energy content of 4.77×10^{-7} J/kg and a density of 6.8×10^2 kg/m^3. Assume that the auto can travel 8.5×10^3 km on a m^3 of gasoline (8.5 km on a liter). [Answer: The cyclist uses about 1.9×10^6 J to travel the 20 km; the car uses 7.5×10^7 J or 40 times more for the same distance.]

PROBLEM 2.6 Suppose that the elevator is broken in the building in which you work and you have to climb 9 stories—a height of 45 m—above ground level. How much extra energy (in kcal and J) will this external work cost if your mass is 70 kg and your body works at 15% efficiency? [Answer: 49 kcal or 2.05×10^5 J]

PROBLEM 2.7 What is the approximate maximum work efficiency of a trained cyclist? How does this compare to the maximum work efficiency of a steam engine?

PROBLEM 2.8 A 70 kg hiker climbs a mountain 1000 m high in 3 hours.
(a) Calculate the external work done by the climber.
[Answer: 6.9×10^5 J]
(b) Assuming the work was done at a steady rate during the 3-hour period, calculate the hiker's average power expenditure during the climb. [Answer: 64 W]
(c) Assuming the average O_2 consumption during the climb was 2×10^{-3} m^3/min (2 liters/min) corresponding to an energy expenditure of 4×10^4 J/min, find the efficiency of the hiker's body. [Answer: 9.4%]
(d) How much energy appeared as heat in the body?
[Answer: 6.6×10^6 J]

2.4 Heat Losses from the Body

Birds and mammals are homeothermic (warm-blooded), while other animals are poikilothermic (cold-blooded). The terms "warm-blooded" and "cold-blooded" are misleading, for a poikilothermic animal such as a frog or a snake will have a higher body temperature on a hot day than a mammal. Birds and mammals both have mechanisms to keep their body temperatures constant despite fluctuations in the environmental temperature. Constant body temperatures permit metabolic processes to proceed at constant rates and these animals to remain active even in cold climates. Birds have a higher body temperature than humans, which helps them radiate heat since they cannot sweat.

Because the body is at a constant temperature, it contains stored heat energy that is essentially constant as long as we are alive. However, when metabolic activity ceases at death, the stored heat is given off at a predictable rate until the body cools to the surrounding temperature. The body temperature of a recently deceased person can thus be used to estimate the time of death.

Although the normal body (core) temperature is often given as 37 C, or 98.6 °F, only a small percentage of people have exactly that temperature. If we measured the temperatures of a large number of healthy people, we would find a distribution of temperatures, with nearly everyone falling within ±0.5 C (~1 °F) of the normal temperature. The rectal temperature is typically 0.5 C (~1 °F) higher than the oral temperature. The temperature depends upon the time of the day, being lower in the morning; the temperature of the environment; and the amount of recent physical activity, the amount of clothing, and the health of the individual. The rectal temperature after hard exercise may be as high as 40 C (104 °F).

Figure 2.4 is a schematic diagram of the body's heating and cooling system. The figure does not show heat transfer by food, drink, and wastes or energy used for external work. Heat is generated in the organs and tissues of the body; most of it is dissipated at the skin's surface. Nearly all heat lost is by radiation, convection, and evaporation (of perspiration). In addition, some cooling of the body takes place when cool inhaled air is heated and moisture is added to the air breathed out. Eating hot or cold food may also heat or cool the body. For the body to hold its temperature close to its normal value it must have a thermostat better than a home thermostat. The hypothalamus of the brain contains the body's thermostat. If the core temperature rises, for example, due

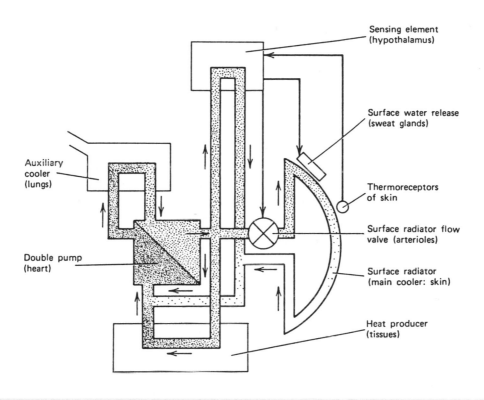

Figure 2.4. Schematic of heat loss system. The density of the dots suggests the heat content of the blood. It is coolest after leaving the skin.

to heavy exertion, the hypothalamus initiates vasodilation, i.e., the blood vessels near the surface expand and carry more blood and heat to the skin. This increases the skin temperature, which promotes sweating and radiation losses (see Section 2.4.1). Both of these reactions increase the heat loss to the environment. If external conditions cause the skin temperature to drop below normal, such as swimming in cold water, thermoreceptors on the skin inform the hypothalamus and shivering is initiated; this increase in involuntary muscle activity is an attempt to increase the core temperature.

The rate of heat production of the body for a 2400 kcal/day diet (assuming no change in body weight) translates to about 1.7 kcal/min or 120 J/s (120 W). If the body is to maintain a constant temperature, it must lose heat at the same rate. The actual amount of heat lost by radiation, convection, evaporation of sweat, and breathing depends on a number of factors: the temperature of the surroundings; the temperature, humidity, and motion of the air. Other factors are the physical activity of the body; the amount of the body exposed; and the amount of insulation on the body (clothes and fat). We now discuss each of the mechanisms of heat loss for the case of a nude body (to simplify so as not to have to include the role of clothing).

2.4.1 Heat Loss by Radiation

All objects, regardless of their temperature, emit energy in the form of electromagnetic radiation. In general, the amount of energy emitted by the body is proportional to the absolute temperature raised to the fourth power (Equation 2.5).

$$E_r = \epsilon \, A \, \sigma \, T^4 \qquad (2.5)$$

where E_r is the rate of energy emitted; ϵ is the emissivity, which has values of $0 \leq \epsilon \leq 1$ and accounts for a surface not being a perfect emitter; A is the area, σ is the Stefan-Boltzmann constant, and T is the absolute temperature. The body also receives radiant energy from surrounding objects. The approximate difference between the energy radiated by the body and the energy absorbed from radiation from the surroundings can be calculated using equation 2.6.

$$H_r = K_r \, A_r \, \epsilon \, (T_s - T_w) \qquad (2.6)$$

where H_r is the rate of energy loss (or gain) due to radiation. A_r is the effective body surface area emitting the radiation, ϵ is the emissivity of the surface, T_s is the skin temperature (C), and T_w is the temperature of the surrounding walls (C). [We can use C here instead of K, as we are computing ΔT. K_r is a constant that depends upon various physical parameters and is about 2.1×10^4 J/(m^2 hr C) (5.0 kcal/(m^2 hr C)]. The emissivity ϵ in the infrared region is independent of the color of the skin and is very nearly equal to 1, indicating that the skin at this wavelength is almost a perfect absorber and emitter of radiation. (If we could see the deep infrared emitted by the body, we would all be "black.")

Under normal conditions about half of our energy loss is the result of radiation, even if the temperature of the surrounding walls is not much lower than body temperature.

2.9

PROBLEM

If a nude body has an effective surface area of 1.2 m² and a skin temperature of 34 C (93.2 °F), compute the rate (in kcal/hr and W) at which it will lose heat to walls maintained at 25 C (77 °F). [Answer: 54 kcal/hr = 62.8 W.] (Note: This amounts to about half of the body's heat loss. Most of the remaining heat loss is due to convection.)

2.4.2 Heat Loss by Convection

The heat loss due to convection (H_c) is given approximately by the equation

$$H_c = K_c A_c (T_s - T_a) \tag{2.7}$$

where K_c is a parameter that depends upon the movement of the air, A_c is the effective surface area, T_s is the temperature of the skin, and T_a is the temperature of the air. When the body is resting and there is no apparent wind, K_c is about 2.3 kcal/(m² hr C).

When the air temperature is 25 C, the skin temperature is 34 C, and the effective surface area is 1.2 m². The nude body loses about 25 kcal/hr by convection. This amounts to about 25% of the body's heat loss. When the air is moving, the constant K_c increases according to the empirical equation

$$K_c = 10.45 - v + 10 \sqrt{v} \ \text{kcal/m}^2/\text{hr C} \tag{2.8}$$

where the wind speed v is in meters per second. This equation is valid for speeds between 2 m/s (~5 mph) and 20 m/s (~45 mph).

Clearly the convective heat loss is greater when the air is moving than when it is still. This has led to the windchill concept. The temperature a person "feels" on a windy day is colder than the measured temperature. The weather forecast in winter often cites two temperatures, one representing the real temperature, while the other represents the effect of the wind and is called the wind chill temperature. Table 2.5 depicts a collection of these two temperatures based on the formula for K_c given above. For example, at an actual temperature of −20 C and a wind speed of 10 m/s (a stiff breeze) the cooling effect on the body is the same as −40 C on a calm day.

Table 2.5. Wind Chill Factor

	Actual Temperature (C)						
	30	20	10	0	−10	−20	−30
Wind Speed) (m/s)	Windchill Temperature (C)						
2	30	20	10	0	−10	−20	−30
5	29	17	5	7	−19	−31	−43
10	29	15	1	−13	−27	−40	−54
15	29	14	−1	−16	−30	−45	−60
20	28	13	−2	−17	−32	−48	−63

2.4.3 Heat Loss by Evaporation

The method of heat loss we are most familiar with is by evaporation. Under normal temperature conditions and in the absence of hard work or exercise, this method of cooling is rather unimportant compared to radiative and convective cooling.

There is some heat loss due to perspiration even when the body does not feel sweaty. It amounts to about 7 kcal/hr, or 7% of the body's heat loss. Under extreme conditions of heat and exercise, an individual may sweat more than 1 liter (10^{-3} m^3) of liquid per hour. Since each kilogram of water that evaporates carries with it the heat of vaporization of 580 kcal, the evaporation of 1 liter (1 kg) carries with it 580 kcal. Of course, the sweat must evaporate from the skin in order to give this cooling effect; sweat that runs off the body provides essentially no cooling. The amount evaporated depends upon the air movement and the relative humidity.

A similar loss of heat is due to the evaporation of moisture in the airways and lungs. When we breathe in air, it becomes saturated with water in the lungs. The additional water in the expired air carries away the same amount of heat as if it were evaporated from the skin. Also, when we inspire cold air, we warm it to body temperature and lose heat. Under typical conditions the total respiratory heat loss is about 14% of the body's heat loss.

2.4.4 Venous Blood Flow Helps Control our Skin Temperature

Since the radiation of heat from the body and the transfer of heat to the air depend upon the skin temperature, any factors that affect the skin temperature also affect the heat loss. The body has the ability to select the path for blood returning from the hands and feet. In cold weather blood is returned to the heart via internal veins that are in contact with the arteries carrying blood to the extremities. In this way some of the heat from the blood going to the extremities is used to heat the returning blood. This counter-current heat exchange lowers the temperature of the extremities and reduces the heat loss to the environment. In the summertime or in a warm environment, the returning venous blood flows near the skin, raising the temperature of the skin and thus increasing the heat loss from the body.

2.4.5 The Effect of Clothing—the Clo

Our discussion of heat loss mechanisms has been restricted to heat loss from the nude body—an interesting, but somewhat uncommon case. Including the insulation of clothing in the heat loss equations makes the calculations more difficult. A typical, comfortable skin temperature is 34 C. This temperature can be maintained by suitably adjusting the clothing to the activity. Studies with clothing have led to the definition of a unit of clothing, the *clo*, which corresponds to the insulating value of clothing needed to maintain a subject in comfort sitting at rest in a room at 21 C (70 °F) with air movement of 0.1 m/s and air humidity of less than 50%. One clo of insulation is equal to a lightweight business suit. Obviously, 2 clos of clothing would enable a person to withstand a colder temperature than 1 clo. Likewise, a person would need a larger clo value to remain comfortable when inactive than when active. It is possible to determine the optimum clothing for comfort under various environmental conditions of temperature, air movement, and humidity for different physical activities. For example, studies show that an individual in the arctic needs clothing with insulation equal to about 4 clos. (Fox fur has an insulating value of about 6 clos.)

2.10

PROBLEM

Consider a person sitting in the nude on a beach in Florida. On a sunny day, radiation energy from the sun is absorbed by the person at the rate of 30 kcal/hr or 34.9 W. The air temperature is a warm 30 C, and the individual's skin temperature is 32 C. The effective body surface area exposed to the sun is 0.9 m^2.

(a) Find the net energy gain or loss from the radiation each hour. [Answer: gain: 21 kcal/hr or 24 W]

(b) If there is a 4 m/s breeze, find the energy lost by convection each hour. [Answer: 48 kcal/hr or 56 W]

(c) If the individual's metabolic rate is 80 kcal/hr (93.0 W), and breathing accounts for a loss of 10 kcal/hr (11.6 W), how much additional heat must be lost by evaporation to keep the body core temperature constant? [Answer: 55.5 kcal/hr or 64.5 W]

2.11

PROBLEM

Compute the rate of convective heat loss by the nude body in still air if the air temperature is 25 C, the skin temperature is 34 C, and the effective surface area is 1.2 m^2. [Answer: 25 kcal/hr or 29 W]

2.12

PROBLEM

(a) Calculate the convective heat loss per hour for a nude standing in a 5 m/s wind. Assume T_s = 33 C, T_a = 20 C, and A_s = 1.2 m^2. [Answer: 767 kcal/hr or 891 W]

(b) For the same wind speed, use Table 2.5 to find the wind chill equivalent temperature. [Answer: 17 C]

2.13

PROBLEM

When an individual is in water, the convective heat loss term is greatly increased. For water immersion, K_c = 16.5 kcal/m^{2}hr C. Assuming the BMR of a resting man is 72 kcal/hr, find the water temperature at which the water heat loss is just balanced by the BMR. Assume A_s = 1.75 m^2 and T_s = 34 C. [Answer: 31.5 C or 88.7 °F.]

3

Muscle and Forces

Physicists recognize four fundamental forces. In the order of their relative strength from weakest to strongest they are: gravitational, electrical, weak nuclear, and strong nuclear. Only the gravitational and electrical forces are of importance in our study of the forces affecting the human body. The electrical force is important at the molecular and cellular levels, e.g., affecting the binding together of our bones and controlling the contraction of our muscles. The gravitational force, though very much weaker than the electrical force by a factor of 10^{39}, is important as a result of the relatively large mass of the human body (at least as compared to its constituent parts, the cells).

3.1 How Forces Affect the Body

We are aware of forces on the body such as the force involved when we bump into objects. We are usually unaware of important forces inside the body, for example, the muscular forces that cause the blood to circulate and the lungs to take in air. A more subtle example is the force that determines if a particular atom or molecule will stay at a given place in the body. For example, in the bones there are many crystals of bone mineral (calcium hydroxyapatite) that require calcium. A calcium atom will become part of the crystal if it gets close to a natural place for calcium and the electrical forces are great enough to trap it. It will stay in that place until local conditions have changed and the electrical forces can no longer hold it in place. This might happen if the bone crystal is destroyed by cancer. We do not attempt to consider all the various forces in the body in this chapter; it would be an impossible task.

Medical specialists who deal with forces are (a) physiatrists (specialists in physical medicine) who use physical methods to diagnose and treat disease, (b) orthopedic specialists who treat and diagnose diseases and abnormalities of the musculoskeletal system, (c) physical therapists, (d) chiropractors who treat the spinal column and nerves, (e) rehabilitation specialists, and (f) orthodontists who deal with prevention and treatment of irregular teeth.

3.1.1 Some Effects of Gravity on the Body

One of the important medical effects of gravity is the formation of varicose veins in the legs as the venous blood travels against the force of gravity on its way to the heart. We discuss varicose veins in Chapter 8, *Physics of the Cardiovascular System*. Yet gravitational force on the skeleton also contributes in some way to healthy bones. When a person becomes "weightless," such as in an orbiting satellite, he or she loses some bone mineral. This may be a serious problem on very long space journeys. Long-term bed rest is similar in that it removes much of the force of body weight from the bones which can lead to serious bone loss.

3.1.2 Electrical Forces in the Body

Control and action of our muscles is electrical. The forces produced by muscles are caused by electrical charges attracting opposite electrical charges. Each of the trillions of living cells in the body has an electrical potential difference across the cell membrane. This is a result of an imbalance of the positively and negatively charged ions on the inside and outside of the cell wall (see Chapter 9, *Electrical Signals from the Body*). The resultant potential difference is about 0.1 V, but because of the very thin cell wall it may produce an electric field as large as 10^7

V/m, an electric field that is much larger than the electric field near a high voltage power line.

Electric eels and some other marine animals are able to add the electrical potential from many cells to produce a stunning voltage of several hundred volts. This special "cell battery" occupies up to 80% of an eel's body length! Since the eel is essentially weightless in the water, it can afford this luxury. Land animals have not developed biological electrical weapons for defense or attack.

In Chapter 9 we discuss the way we get information about body function by observing the electrical potentials generated by the various organs and tissues.

3.2 Frictional Forces

Friction and the energy loss resulting from friction appear everywhere in our everyday life. Friction limits the efficiency of machines such as electrical generators and automobiles. On the other hand, we make use of friction when our hands grip a rope, when we walk or run, and in devices such as automobile brakes.

Some diseases of the body, such as arthritis, increase the friction in bone joints. Friction plays an important role when a person is walking. A force is transmitted from the foot to the ground as the heel touches the ground (Fig. 3.1a). This force can be resolved into vertical and horizontal components. The vertical reaction force, supplied by the surface, is labeled N (a force perpendicular to the surface). The horizontal reaction component, F_H, must be supplied by frictional forces. The maximum force of friction F_f is usually described by:

$$F_f = \mu N$$

where N is a normal force and μ is the coefficient of friction between the two surfaces. The value of μ depends upon the two materials in contact, and it is essentially independent of the surface area. Table 3.1 gives values of μ for a number of different materials.

The horizontal force component of the heel as it strikes the ground when a person is walking (Fig. 3.1a) has been measured, and found to be approximately 0.15W, where W is the person's weight. This is how large the frictional force must be in order to prevent the heel from slipping. If we let N≈W, we can apply a frictional force as large as $f = \mu W$.

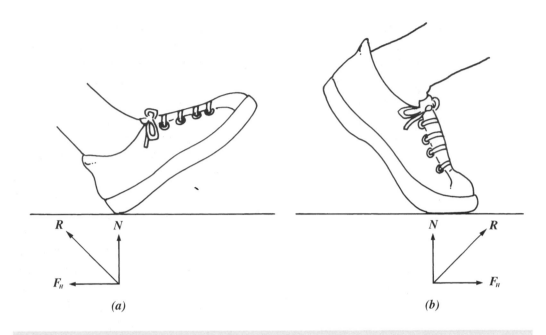

Figure 3.1. Normal walking. (a) Both a horizontal frictional component of force, F_H, and a vertical component of force N with resultant R exist on the heel as it strikes the ground, decelerating the foot and body. The friction between the heel and surface prevents the foot from slipping forward. (b) When the foot leaves the ground, the frictional component of force, F_H, prevents the foot from slipping backward and provides the force to accelerate the body forward. (Adapted from M. Williams and H. R. Lissner, *Biomechanics of Human Motion*, Philadelphia, W. B. Saunders Company, 1962, p. 122, by permission.)

Table 3.1. Examples of Values of Coefficients of Friction

Material	μ (Static Friction)
Steel on steel	0.15
Rubber tire on dry concrete road	1.00
Rubber tire on wet concrete road	0.7
Steel on ice	0.03
Between tendon and sheath	0.013
Normal bone joint	0.003

For a rubber heel on a dry concrete surface, the maximum frictional force can be as large as $f \cong W$, which is much larger than the needed horizontal force component (0.15W). In general, the frictional force is large enough both when the heel touches down and when the toe leaves the surface to prevent a person from slipping (Fig. 3.1). Occasionally, a person slips on an icy, wet, or oily surface where μ is less than 0.15. This is not only embarrassing; it may result in broken bones. Slipping can be minimized by taking very small steps.

Friction must be overcome when joints move, but for normal joints it is very small. The coefficient of friction in bone joints is usually much lower than in engineering-type materials (Table 3.1). If a disease of the joint exists, the friction may become significant. Synovial fluid in the joint is involved in lubrication, but controversy still exists as to its exact behavior. Joint lubrication is considered further in Chapter 4.

The saliva we add when we chew food acts as a lubricant. If you swallow a piece of dry toast you become painfully aware of this lack of lubricant. Most of the large internal organs in the body are in more or less constant motion and require lubrication. Each time the heart beats, it moves. The lungs move inside the chest with each breath, and the intestines have a slow rhythmic motion (peristalsis) as they move food toward its final destination. All of these organs are lubricated by a slippery mucus covering to minimize friction.

3.3 Forces, Muscles, and Joints

In this section we discuss forces in the body and forces at selected joints and give some examples of muscle connections to tendons and bones of the skeleton. Since movement and life itself depends critically on muscle contraction, we start by examining muscles.

3.3.1 Muscles and Their Classification

Several schemes exist to classify muscles. One widely used approach is to describe how the muscles appear under a light microscope. Skeletal muscles have small fibers with alternating dark and light bands, called *striations*—hence the name *striated muscle*. The fibers are smaller in diameter than a human hair and can be several centimeters long. The other muscle form, which does not exhibit striations, is called *smooth muscle*.

The fibers in the striated muscles connect to tendons and form bundles. Good examples are the biceps and triceps muscles depicted in Fig. 3.2, which will be examined further later in this section.

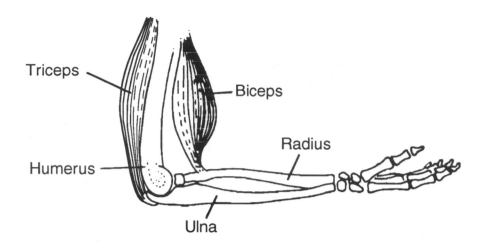

Figure 3.2. Schematic view of the muscle system used to bend the elbow. Biceps bend the elbow to lift, triceps straighten it.

Closer examination of the fibers show still smaller strands called *myofibrils* that, when examined by an electron microscope, consist of even smaller structures called *filaments*. The latter are composed of proteins. As shown schematically in Fig. 3.3, the filaments appear in two forms: (1) thick filaments that are composed of the protein myosin and are about 10 nm in diameter and 2000 nm (2×10^{-6} m or 2 micrometers) long, and (2) thin filaments that are composed of the protein actin and are about 5 nm in diameter and 1500 nm long. During contraction, an electrostatic force of attraction between the bands causes them to slide together, thus shortening the overall length of the bundle. A contraction of 15–20% of their resting length can be achieved in this way. The contraction mechanism at this level is not completely understood. It is evident that electrical forces are involved, as they are the only known force available. It should be emphasized that muscles produce a force only in contraction, that is, during a shortening of the muscle bundle.

Smooth muscles do not form fibers and, in general, are much shorter than striated muscles. Their contraction mechanism is different, and in some cases they may contract more than the resting length of an individual muscle cell. This effect is believed to be caused by the slipping of muscle cells over each other. Examples of smooth muscles in the body are circular (sphincter) muscles around the anus, bladder, and intestines, and in the walls of arteries and arterioles (where they control blood flow).

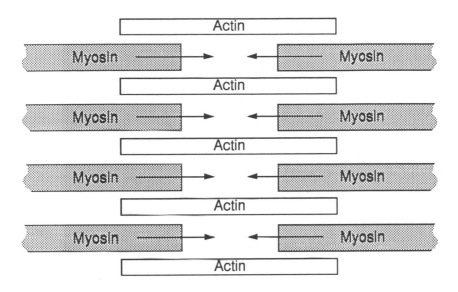

Figure 3.3. Schematic view of actin and myosin filaments with arrows showing the sliding movement between the filaments associated with muscle contraction.

Sometimes muscles are classified as to whether their control is voluntary (generally, the striated muscles) or involuntary (generally, the smooth muscles). This classification breaks down, however; the bladder has smooth muscle around it, yet is (usually) under voluntary control.

A third method of classifying muscles is based on the speed of the muscle's response to a stimulus. Striated muscles usually contract in times around 0.1 s (for example, the time to bend an arm), while smooth muscles may take several seconds to contract (control of the bladder).

3.3.2 Muscle Forces Involving Levers

For the body to be at rest and in equilibrium (static), the sum of the forces acting on it in any direction and the sum of the torques about any axis must both equal zero.

Many of the muscle and bone systems of the body act as levers. Levers are classified as first-, second-, and third-class systems (Fig. 3.4). Third-class levers are most common in the body, while first-class levers are least common.

Third-class levers, however, are not very common in engineering. To illustrate why this is so, suppose you were to open a door whose

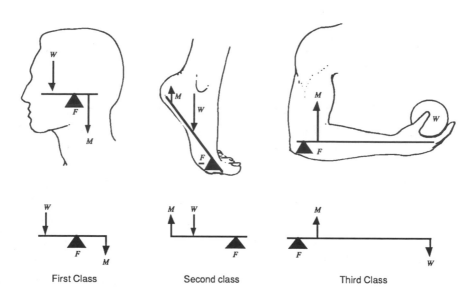

Figure 3.4. The three lever classes in the body and schematic examples of each. W is a force that is usually the weight, F is the force at the fulcrum point, and M is the muscular force. Note that the different levers depend upon different arrangement of the three forces, M, W, and F.

doorknob was located close to the hinge side of the door. It requires a certain amount of torque to open the door. Recall that torque is the product of the applied force and a lever arm that describes the effect this force will have to produce rotation about the hinge. Since the lever arm in this example is small, it follows that it will require a great deal of force to open the door. Finally, note that the applied force in this example must move the door near the hinge only a short distance to open the door. In the case of humans, this type of lever system amplifies the motion of our limited muscle contraction and thus allows for larger (and faster!) movement of the extremities. We give an example of movement of the forearm later in this section.

Muscles taper on both ends where tendons are formed. Tendons connect the muscles to the bones. Muscles with two tendons on one end are called biceps; those with three tendons on one end are called triceps. Because muscles can only contract, muscle groups occur in pairs; one group serves to produce motion in one direction about a hinged joint, and the opposing group produces motion in the opposite direction. The rotation of the forearm about the elbow is an excellent example of this principle. The biceps act to raise the forearm toward the upper arm, while the triceps (on the back of the upper arm) pull the forearm away from the upper arm. Try this yourself a few times, feeling the action of these upper arm muscles with your other hand.

Try the following to experience the advantages and disadvantages of a third-class lever system. Place a large plastic bucket on a table and load it with two 5 kg masses (weight, 98 N or 22 lb). Wrap the handle of the bucket with a cloth to provide a softer suspension point. Lift the bucket with one hand, keeping the angle between your forearm and upper arm about 90°. Now repeat the experiment of lifting the bucket with the handle further up your forearm, say halfway to the elbow. Can you feel the difference in the force required in your biceps? By how much has it changed—by sense and by calculation (see below)? Repeat this experiment with varying angles between the two parts of your arm.

Let's consider further the case of the biceps muscle and the radius bone acting to support a weight W in the hand (Fig. 3.5a). Figure 3.5b shows the forces and dimensions of a typical arm. We can find the force supplied by the biceps if we sum the torques (force times distance—moment arm) about the pivot point at the joint. There are only two torques: that due to the weight W (which is equal to 30W acting clockwise) and that produced by the muscle force M (which acts counterclockwise and of magnitude 4M). With the arm in equilibrium 4 M must equal 30 W, or 4 M − 30 W = 0 and M = 7.5 W. Thus, a muscle force 7.5 times the weight is needed. For a 100 N (~22 lb) weight, the muscle force is 750 N (~165 lb).

For individuals building their muscles through weight lifting, the exercise of lifting a dumbbell as in Fig. 3.5 is called a dumbbell curl. A trained individual could probably curl about 200 N (~44 lb) requiring the biceps to provide 1500 N (~330 lb) force.

In our simplification of the example in Fig. 3.5b, we neglected the weight of the forearm and hand. This weight is not present at a particular point but is nonuniformly distributed over the whole forearm and hand. We can imagine this contribution as broken up into small segments and include the torque from each of the segments. A better method is to find the center of gravity for the weight of the forearm and hand and assume all the weight is at that point. Figure 3.5c shows a more correct representation of the problem with the weight of the forearm and hand, H, included. A typical value of H is 15 N (~3.3 lb). By summing the

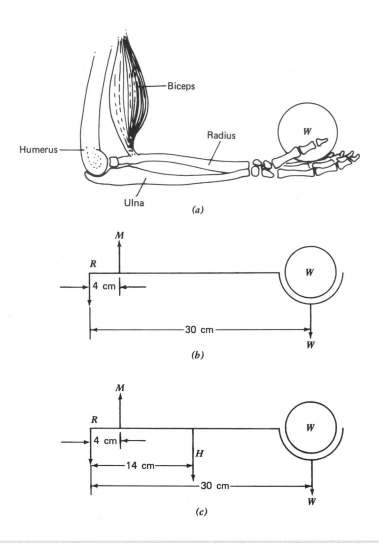

Figure 3.5. The forearm. (a) The muscle and bone system. (b) The forces and dimensions: R is the reaction force of the humerus on the ulna, M is the muscle force supplied by the biceps, and W is the weight in the hand. (c) The forces and dimensions where the weight of the tissue and bones of the hand and forearm H is included. These forces are located at their center of gravity.

torques about the joint we obtain 4 M = 14 H + 30 W, which simplifies to M = 3.5 H + 7.5 W. This simply means that the force supplied by the biceps muscle must be larger than that indicated by our first calculation by an amount 3.5 H = (3.5)(15) = 52.5 N (~12 lb).

What muscle force is needed if the angle of the arm changes from the 90° (between forearm and upper arm) that we have been considering so far, as illustrated in Fig. 3.6a? Figure 3.6b shows the forces we must consider for an arbitrary angle α. If we take the torques about the joint we find that M remains constant as alpha changes! (As you will see if you perform the calculation, this is because the same trigonometric function of α appears in each term of the torque equation.) However, the length of the biceps muscle changes with the angle. Muscle has a minimum length to which it can be contracted and a maximum length to which it can be stretched and still function. At these two extremes, the force the muscle can exert is much smaller. At some point in between, the muscle produces its maximum force (see Fig. 3.7). If the biceps pulls vertically (which is an approximation), the angle of the forearm does not affect the force required; but it does affect the length of the biceps muscle, which in turn affects the ability of the muscle to provide the needed force. Most of us become aware of the limitations of the biceps if we try to chin ourselves. With our arms fully extended we have difficulty, and as the chin approaches the bar the shortened muscle loses its ability to shorten further.

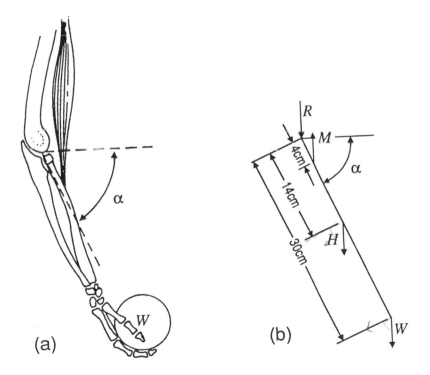

Figure 3.6. The forearm at an angle α to the horizontal. (a) The muscle and bone system. (b) The forces and dimensions.

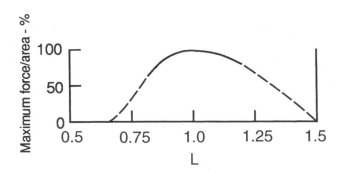

Figure 3.7. At its resting length L a muscle is close to its optimum length for producing force. At about 80% of this length it cannot shorten much more and the force it can produce drops significantly. The same is true for stretching of the muscle to about 20% greater than its natural length. A very large stretch of about 2L produces irreversible tearing of the muscle.

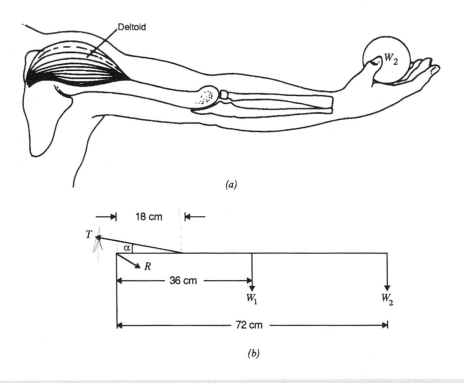

Figure 3.8. Raising the right arm. (a) The deltoid muscle and bones involved. (b) The forces on the arm. T is the tension in the deltoid muscle fixed at the angle α, R is the reaction force on the shoulder joint, W_1 is the weight of the arm located at its center of gravity, and W_2 is the weight in the hand. (Adapted from L. A. Strait, V. T. Inman, and H. J. Ralston, *Amer. J. Phys.*, 15, 1947, p. 379.)

The arm can be raised and held out horizontally from the shoulder by the deltoid muscle (Fig. 3.8a); we can show the forces schematically (Fig. 3.8b). By taking the sum of the torques about the shoulder joint, the tension T can be calculated from:

$$T = (2\,W_1 + 4\,W_2)/ \sin \alpha \qquad (3.1)$$

If $\alpha = 16°$, the weight of the arm $W_1 = 68$ N (~15 lb), and the weight in the hand $W_2 = 45$N (~10 lb), then $T = 1145$ N (~250 lb). The force needed to hold up the arm is surprisingly large.

3.2 PROBLEM

In the lever of the foot shown in Fig. 3.4, is M greater or smaller than the weight on the foot? (Hint: The muscle that produces M is attached to the tibia, a bone in the lower leg.)

3.3 PROBLEM

Show that for Fig. 3.6, the muscle force is independent of the angle.

3.4 PROBLEM

Derive Equation 3.1 for the arm and deltoid muscle system.

3.5

PROBLEM

It is known that the human biceps can produce a force of approximately 2600 N. Why can't you pick up an object with your hand which weighs 2600 N?

3.6

PROBLEM

If you turn your hand over and press it against a table, you have a first class lever system (see sketch). In this case, the biceps muscle group is relaxed and is ignored. The force of the hand F on the table is balanced by the force supplied by the triceps M pulling on the ulna and the fulcrum force R located where the humerus makes contact with the ulna. For the parameters shown below and for a force F = 100 N (22 lb), find the force needed from the triceps. Ignore the mass of the arm and hand.

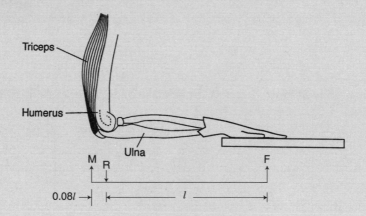

3.7

PROBLEM

One first-class lever system involves the extensor muscle, which exerts a force M to hold the head erect; the force W of the weight of the head, acting at its center of gravity (cg), lies forward of the force F exerted by the first cervical vertebra (see sketch on the next page). The head has a mass of about 3 kg, or weight $W \cong 30$ N.

(a) Find F and M. [Answer: F = 48 N; M = 18 N]
(b) If the area of the first cervical vertebra, on which the head rests, is 5×10^{-4} m^2, find the stress (force per unit area: N/m^2) on it. [Answer: 9.6×10^4 N/m^2]

(c) How does this stress compare with the rupture compression strength for vertebral disks (1.1×10^7 N/m^2)? [Answer: 1.3×10^6 N/m^2]

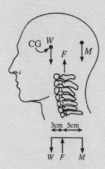

3.3.3 The Spinal Column

Bones provide the main structural support for the body (see Chapter 4, Fig. 4.1). Examination of that figure shows that the cross-sectional area of the supporting bones generally increases from head to toe. These bones provide the support for the additional weight of muscle and tissue as one moves downward to the soles of the feet. The body follows the same engineering principles as used in the design of a building where the major support strength is in the base. (Note, however, that there are exceptions; the femur is larger than the tibia and fibula, the supporting bones in the legs.)

Load-bearing bones are optimized for their supporting tasks. The outside or compact dense bone is designed to carry compressive loads. The inner spongy or cancellous bone, at the ends of long bones and in the vertebrae, has thread-like filaments of bone (trabeculae) which provide strength yet are light in weight. Engineering examples of such construction would be honeycomb structures used to strengthen aircraft wings, the use of lightweight graphite fibers in composite materials, and the framework used to support and strengthen buildings.

The vertebrae are examples of load-bearing bones. The spinal column of a skeleton is shown in Fig. 3.9. Note that the vertebrae increase in both thickness and cross-sectional area as you go from the neck (cervical) region to the lower back (lumbar) region. A larger surface area is needed to support the additional body mass above each succeeding vertebra. There are fibrous discs between the vertebrae that cushion the downward forces and other impacts on the spinal column. However, the pressure (force/area) remains approximately constant for all discs. The discs rupture at a stress (pressure) of about 10^7 N/m^2 (10^7 Pa; 100 atmospheres).

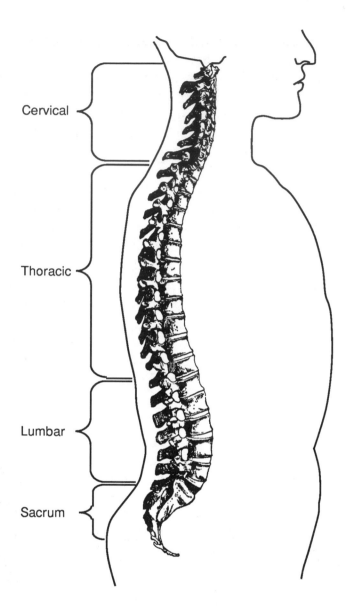

Figure 3.9. The spinal column provides the main support for the head and trunk of the body. The column has an "S" shape, and the vertebrae increase in cross-sectional area as the supporting load increases. The length of the column for a typical adult male is about 0.7 m.

The length of the spinal column shortens slightly from its normal length of about 0.7 m (male) by as much as 0.015 m (1.5 cm = 0.6 in) after arising from sleep. The original length is restored after a night's sleep. However, the spinal column does shorten permanently with age most often as the result of osteoporosis and compression of the discs, which is particularly common in elderly women. Osteoporosis causes bone to weaken and eventually to collapse. This is discussed further in the next chapter.

The spinal column has a normal curvature for stability. Viewed from the right side the lower portion of the spine is shaped like a letter "S" as shown in Fig. 3.9. Lordosis, kyphosis, and scoliosis are deviations in the shape of the spine. *Lordosis*, too much curvature, often occurs in the lumbar region. A person with this condition is sometimes called sway-backed (Fig. 3.10a). *Kyphosis* is an irregular curvature of the spinal column as seen from the side; frequently it leads to a hump in the back. A person with this condition is often referred to as hunch-backed (Fig. 3.10b). *Scoliosis* is a condition in which the spine curves in an "S" shape as seen from the back (Fig. 3.10c). Normal posture is shown in Fig. 3.10d.

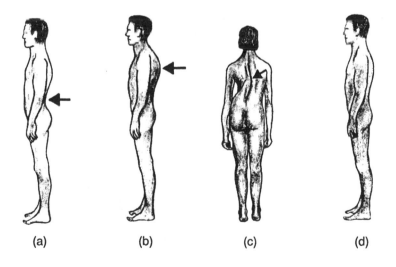

(a)   (b)   (c)   (d)

Figure 3.10. Sketches for the abnormal spinal conditions of (a) lordosis (or sway-back), (b) kyphosis (or hunch-backed), and (c) scoliosis. (d) The normal condition. (Adapted from *A Guide to Physical Examination*, B. Bates, J. P. Lippincott, Philadelphia, PA, pp. 261–261, (1974) by permission.)

PROBLEM

3.8

The discs in the spinal column can withstand a stress (force per unit area) of 1.1×10^7 N/m^2 before they rupture.

(a) If the cross-sectional area of your discs is 10 cm^2, what is the maximum force that can be applied before rupture takes place? [Answer: 1.1×10^4 N]

(b) Estimate the stress at a disc located at the level of the center of gravity of your body when you are standing vertically. [Answer: 3.5×10^5 N/m^2]

(c) What types of situations might the body experience where the stress on this vertebra would be much larger than in (b) above?

3.3.4 Stability While Standing

In an erect human viewed from the back, the center of gravity (cg) is located in the pelvis in front of the upper part of the sacrum at about 58% of the person's height above the floor. A vertical line from the cg passes between the feet. Poor muscle control, accidents, disease, pregnancies, overweight conditions, or poor posture change the position of the cg to an unnatural location in the body as illustrated in Fig. 3.11. An overweight condition (or a pronounced slump) lead to a forward shift of the cg, moving the vertical projection of it under the balls of the feet where the balance is less stable. The person may compensate by tipping slightly backward.

To retain stability while standing, you have to keep the vertical projection of your cg inside the area covered by your feet (Fig. 3.12a). If the vertical projection of your cg falls outside this area, you will tip over. When your feet are close together (Fig 3.12a) you are less stable than when they are spread apart (Fig 3.12b). Likewise, if the cg is lowered, you become more stable. A cane or crutch also improves your stability (Fig. 3.12c). Comparing the stability of a human with a four-legged animal, it is clear that the animal is more stable because the area between its four feet is larger than for two-legged humans. Thus it is understandable that a human baby takes about ten months before it is able to stand while a newborn four-legged animal achieves this in less than two days (in the wild, less than one hour), a useful condition for survival.

The body compensates its stance when lifting a heavy suitcase with one arm. The opposite arm moves out and the body tips away from the object to keep the cg properly placed for balance. (Try lifting the bucket used in Problem 3.1 out to the side to see how this works.) People who

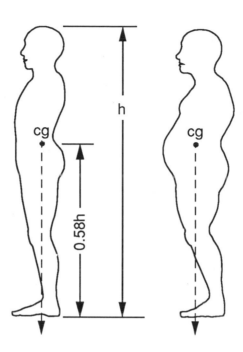

Figure 3.11. (a) The center of gravity of a normal person is located about 58% of the person's height above the soles of their feet. (b) An overweight condition can shift the cg forward so that the vertical projection of it passes underneath the balls of the feet, causing the body to compensate by assuming an unnatural position leading to possible muscle strain. (After C. R. Nave and B. C. Nave, *Physics for the Health Sciences*, W. B. Saunders Company, 1975, p. 24 by permission.)

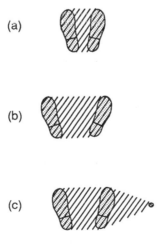

Figure 3.12. The body remains stable as long as the vertical projection of the cg remains inside the cross-hatched area between the feet. (a) The stable area when the feet are close together, (b) the stable area when the feet are spread apart, and (c) the stable area when a cane or crutch is used.

have had an arm amputated are in a situation similar to a person carrying a suitcase. They compensate for the weight of the remaining arm
by bending the torso; however, continued bending of the torso leads to
spine curvature. A common prosthesis is an artificial arm with a mass
equal to the missing arm. Even though the false arm may not function,
it helps to prevent distortion of the spine.

3.3.5 Lifting and Squatting

The spinal cord is enclosed and protected by the spinal column. The spinal
cord provides the main pathway for the transmission of nerve signals to
and from the brain. The discs separating the vertebrae can be damaged;
one common back ailment is called a slipped disc. The condition occurs
when the wall of the disc weakens and tears, leading to a bulge that sometimes pushes against nerves passing through the special holes (foramina)
on the sides of each vertebra. Extended bed rest, traction, physical ther-
apies, and surgery are all used to alleviate this condition.

An often abused part of the body is the lumbar (lower back)
region, shown schematically in Fig. 3.13. Lumbar vertebrae are subject to very large forces—those resulting from the weight of the body
and also the forces you create in the lumbar region by lifting. The
figure illustrates the large compressive force (labeled R) on the fifth
lumbar vertebra (labeled L5). When the body is bent forward at 60°
to the vertical and there is a weight of 225 N (~50 lb) in the hands,
the compressive force R can approach 3800 N (~850 lb, or about six
times an average body weight).

It is not surprising that lifting heavy objects incorrectly is a primary
cause of low back pain. Since low back pain can be serious and is not
well understood, physiologists are interested in finding out exactly how
large the forces are in the lumbar region. Measurements of pressure in
the discs have been made by inserting a hollow needle connected to a
calibrated pressure transducer into the gelatinous center of an interver-
tebral disc. This device measures the pressure within the disc. The
pressures in the third lumbar disc for an adult in different positions are
shown in Fig. 3.14a and 3.14b. Even when standing erect there is a relatively large pressure in the disc as a result of the combined effects of
weight and muscular tension. If the disc is overloaded as might occur in
improper lifting, it can rupture (or slip), causing pain either from the rupture or by allowing irritating materials from inside the disc to leak out.

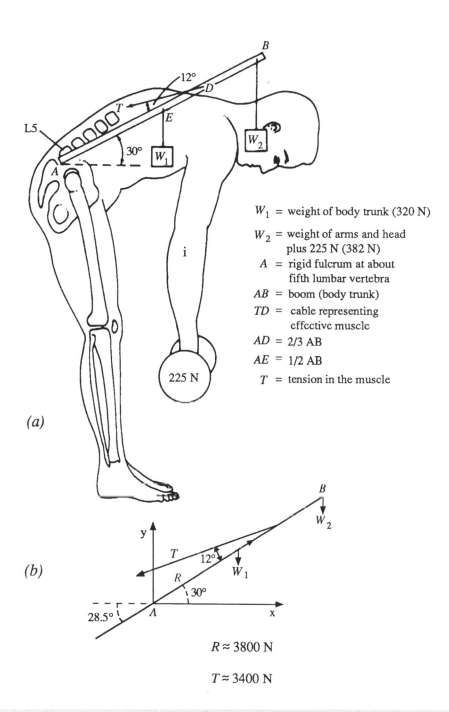

W_1 = weight of body trunk (320 N)

W_2 = weight of arms and head plus 225 N (382 N)

A = rigid fulcrum at about fifth lumbar vertebra

AB = boom (body trunk)

TD = cable representing effective muscle

AD = 2/3 AB

AE = 1/2 AB

T = tension in the muscle

(a)

(b)

$R \approx 3800$ N

$T \approx 3400$ N

Figure 3.13. Lifting a weight. (a) Schematic of forces used. (b) The forces where T is an approximation for all of the muscle forces and R is the resultant force on the fifth lumbar vertebra (L5). Note that the reaction force R at the fifth lumbar vertebra is large. (Adapted from L. A. Strait, V. T. Inman, and H. J. Ralston, *Amer. J. Phys.*, 15, 1947, pp. 377–378.)

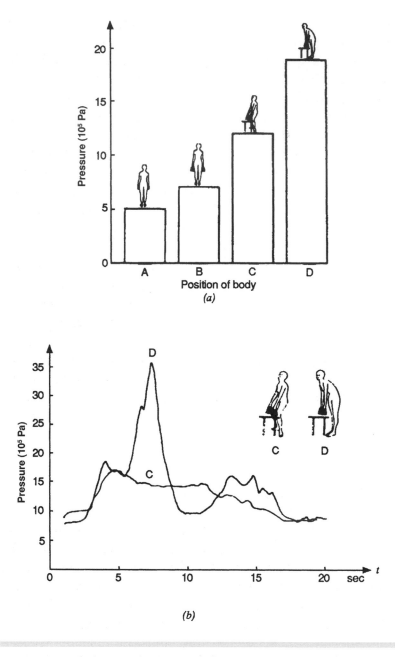

Figure 3.14. Pressure on the spinal column. (a) The pressure on the third lumbar disc for a subject (A) standing, (B) standing and holding 20 kg, (C) picking up 20 kg correctly by bending the knees, and (D) picking up 20 kg incorrectly without bending the knees. (b) The instantaneous pressure in the third lumbar disc while picking up and replacing 20 kg correctly and incorrectly. Note the much larger peak pressure during incorrect lifting. (Adapted from A. Nachemson and G. Elfstrom, *Scand. J. Rehab. Med.*, Suppl. 1, 1970, pp. 21–22.)

It has been argued that low back pain is the price that humans pay for being erect; however, disc degeneration also occurs in four-legged animals (in particular, in dachshunds). Disc failures for both animals and humans occur in regions under the greatest stress.

Just as forces can be transmitted over distances and around corners by cable and pulley systems, the forces of muscles in the body are transmitted by tendons. Tendons, the fibrous cords which connect the muscle end to a bone, minimize the bulk present at a joint. For example, the muscles that move the fingers to grip objects are located in the forearm, and long tendons are connected to appropriate places on the finger bones. Of course, the tendons have to remain in their proper locations to function properly. Arthritis in the hands often prevents the tendons from fully opening and closing the hands.

In the leg, a tendon passes over a groove in the kneecap (patella) and connects to the shin bone (tibia). With your leg extended you can move the patella with your hand but with your knee flexed you cannot; the patella is held rigidly in place by the force from the tendon as shown in Fig. 3.15. The patella also serves as a pulley for changing the direction of the force. This also acts to increase the mechanical advantage of the muscles that straighten the leg. Some of the largest forces in the body occur at the patella. When you are in a deep squatting position, the tension in the tendons that pass over the patella may be more than two times your weight (Fig. 3.15).

3.3.6 Forces on the Hip and Thigh

When you are walking, there is an instant when only one foot is on the ground and the cg of your body is directly over that foot. Fig. 3.16a shows the forces acting on that leg. These forces are (1) the upward vertical force on the foot, equal to the weight of the body, W; (2) the weight of the leg, W_L, which is approximately equal to W/7; (3) R, the reaction force acting between the hip and the femur; and (4) the tension, T, in the muscle group between the hip and the greater trochanter on the femur. The latter provides the force to keep the body in balance.

The various dimensions and the angle shown in Fig. 3.16 have been taken from cadaver measurements. Solving the equations for equilibrium in this example, it is found that T = 1.6 W and R = 2.4 W at the hip joint. Thus for a 70 kg individual, the head of the femur experiences a force of over 1600 N ($\approx$350 lb) or 2.4 times the body weight!

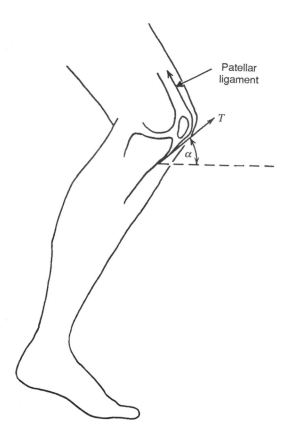

Figure 3.15. Diagram of the tensile force on the patellar ligament during squatting. The tension T is very large when a person is in a low squat.

When there is injury to the muscle group at the hip, or damage to the hip joint, the body reacts by trying to reduce the forces that cause pain—T and R in Fig. 3.16a. It does this by tipping the body so that the cg is directly over the ball of the femur and the foot (Fig. 3.16b). This reduces the muscle force, T, to nearly zero, and the reaction force, R, becomes approximately the body weight W minus one leg, or (6/7) W. R is now pointing directly downward. This reduces the forces T and R by a large amount and helps the healing process. However, the downward reaction force causes the head of the femur to grow upward, while the ball of the femur on the other leg does not change. Eventually this leads to uneven growth at the hip joints and possible permanent curvature of the spine.

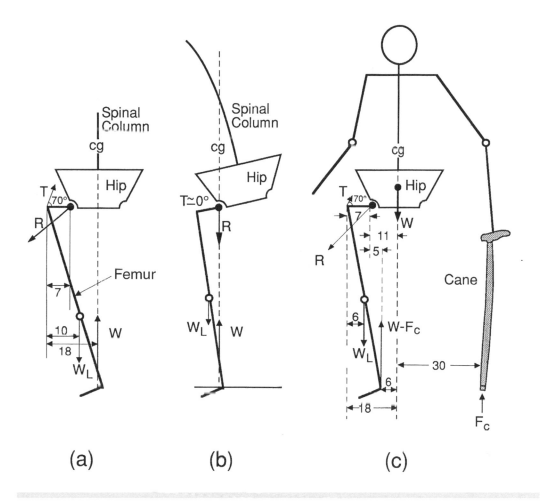

Figure 3.16. A diagram that shows approximately the forces and dimensions (in cm) for the hip-leg under different conditions. (a) When the person is standing on one foot. The vertical upward force on the foot is the person's weight, W. The weight of the leg, W_L, is taken to be W/7 and the angle of the hip abductor muscles indicated by T is taken to be 70°. R is the reaction force between the hip and the head of the femur (hip joint). (b) When either the hip joint or abductor muscle is injured, the body is bent to place the cg directly over the ball of the femur and the center of the foot, thus reducing the reaction force, R, and the force of the abductor muscle, T. (c) When a cane is used, the abductor force, T, and the reaction force, R, at the head of the femur are greatly reduced. The upward force of F_C = W/6 gives T ≈ 0.65W and R ≈ 1.3W, a substantial reduction from that of part (a). (Adapted from M. Williams and H. R. Lissner, *Biomechanics of Human Motion*, Philadelphia, W. B. Saunders Company, 1962, p.110 and from G. B. Benedek and F. M. H. Vilars, *Physics with Illustrative Examples from Medicine and Biology, Vol. 1, Mechanics*, Addison-Wesley, 1973.)

The use of crutches or a cane reduces the force on the hip joint. The physics of the use of a cane is shown schematically in Fig. 3.16c. There are three forces acting on the body: the weight, W, the force, F_C, pushing upward on the cane, and the upward force on the foot equal to $W - F_C$. Note that the cane is in the hand opposite to the injured hip. Without the cane, we found T = 1.6 W and R = 2.4 W. The use of the cane reduces these forces by allowing the foot to move from the position under the centerline of the body, as in Fig. 3.16a, to a new location closer to being under the head of the femur. The spine is not twisted as it is in Fig. 3.16b. The cane is located 0.3 m from the vertical projection line of the cg. We assume that the cane supports about 1/6 of the body's weight. For the conditions given in Fig. 3.16c, we find T = 0.65 W and R = 1.3 W. Although human nature leads us to hide our handicaps, the use of a cane can considerably aid in the healing process for hip joints.

3.9

PROBLEM

Use the equations of static equilibrium to calculate the forces T and R for the case shown in Fig. 3.16a. [Answer: T = 1.6 W; R = 2.4 W]

3.4 Forces During Collisions

When a portion of the body (or the whole body) bumps into a solid object, it rapidly decelerates, resulting in large forces. If we consider the deceleration to be constant and limit ourselves to one-dimensional motion, we can use the original form of Newton's second law. Force equals the rate of change of momentum. The more common form, mass times acceleration, can be written as:

$$F = ma = m(\Delta v/\Delta t) = \Delta(mv)/\Delta t$$

or F = the rate of change of momentum.

Newton originally wrote his second law in this form.

3.4.1 Examples of Forces during Collisions

The following example illustrates how this form of Newton's second law can be used to estimate the forces on the body when it collides with something:

Example: A person walking at 1 m/s accidentally bumps her or his head against an overhanging steel beam (ouch!). Assume that the head stops in about $\Delta t = 0.01$ s while traveling an additional distance of 0.005 m (5 mm). The mass of the head is 3 kg. What is the force which caused this deceleration?

> **Answer:** The change of momentum is $\Delta(mv)$ = (3 kg)(0 m/s) − (3 kg)(1 m/s) = −3 kg m/s (the minus sign means that the momentum of the head has decreased; the force is in the opposite direction from the motion). Thus F = (−3 kg m/s)/(0.01 s) = −300 N (about 67 lb force).

Example: If we repeat this accident, with a steel beam with 0.02 m (2 cm) of padding, the time of deceleration is increased to Δt = 0.04 s. What force acts to decelerate the head under these conditions?

> **Answer:** $F = \Delta(mv)/\Delta t$ = (3 kg m/s)/(0.04 s) = 75 N (about 15 lb), a considerable reduction from the first case.

An example of a dynamic force in the body is the apparent increase of weight when the heart beats (systole). About 0.06 kg of blood is given a velocity of about 1 m/s upward in a time of t = 0.1s. The upward momentum given to the mass of blood is (0.06 kg) (1 m/s) = 0.06 kg m/s; thus the reaction force to this movement of the blood is (0.06 kg m/sec)/(0.1 s) or 0.6 N (~0.125 lb, or 2 oz). This is enough to produce a noticeable jiggle on a sensitive spring-type scale (as noted in Chapter 1, *Terminology, Modeling, and Measurement*).

If you jump from a height of 1 m and land stiff-legged, you are in for a shock. Under these conditions, the deceleration of the body takes place mostly through compression of the padding of the feet. We can calculate that the body is traveling at 4.5 m/s (16 km/hr) just prior to hitting; and if the padding collapses by 1 cm, the body stops in about 0.005 s (5 ms). Under these conditions, the force in your legs is almost 100 times your weight (that is, 100 g; see Fig. 3.17). If you land on a

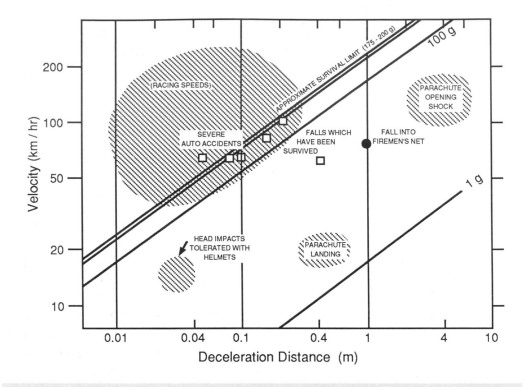

Figure 3.17. A compilation of documented cases of impact results on humans shown as a log-log plot of the velocity on impact versus the deceleration distance during impact. The diagonal lines show the deceleration in terms of acceleration of gravity, g. (One g times your body's mass is equal to your body weight.) The hollow squares represent data from documented free-fall survivors. The shaded areas represent guestimates for the other situations. (After R. G. Snyder, *Bioastronautics Data Book*, Second ed., 1973, p. 228.)

gym mat, the deceleration time would be longer; and if you follow the normal body reaction, you will land on your toes first and bend your knees to decelerate over a much longer time, thus decreasing the landing force.

A current popular form of entertainment is bungee jumping, in which a person is attached to a very stretchable bungee cord and jumps from a considerable height. The bungee cord decelerates the person over a long distance. The thrill comes from the freefall and deceleration. In terms used in Fig. 3.17, the deceleration distances would usually be more than 10 m and the velocities below 100 km/hr. This puts the conditions beyond the upper right region of the figure.

3.10

PROBLEM

A 50 kg person jumping from a height of 1 m is traveling at 4.4 m/s just prior to landing. Suppose the person lands on a pad and stops in 0.2 s. What maximum decelerating force will be experienced? [Answer: F_{max} = 1100 N]

3.4.2 Surviving Falls from Great Heights

You might think that if you jump or fall from a great height your chance of surviving is zero, unless of course you land on something like a giant airbag. In real life, your chances are very small, but not zero. People have survived falls from great heights. It all depends on where and how you land! If you fall on bushes, tree branches, deep snow, or land on the side of a hill, the deceleration forces you experience may be small enough that you could survive. A summary of the hazardous ranges for impact collisions is shown in Fig. 3.17 along with some documented cases. This figure shows the velocity at the time of impact plotted versus the distance needed to stop. One could equally well plot the velocity versus the time needed to stop, but usually the distance is more easily measured. The heavy diagonal lines in the figure indicate the decelerations in terms of the units of gravity, $g = 9.8$ m/s^{-2}. For example, a deceleration of 10 g corresponds to a decelerative force equal to ten times the weight of the object. The double line in the figure represents an estimate of the limit of survivability.

3.4.3 Collisions Involving Vehicles

Collisions of high velocity, modern cars subject occupants to very large accelerative or decelerative forces. The results of these forces on the driver and passengers can be broken bones, internal injuries, and death.

In the 1960s a federally mandated safety program for the automobile was begun. Even earlier, the military, NASA, and scientific groups were studying the forces that the body could withstand. For small controlled forces, this study was conducted using human volunteers. For more extreme limits, cadavers, dummies, or animals were used to determine the tolerance ranges.

Consider a head-on collision with a solid barrier, one of the most serious types of automobile accident. What happens to the automobile and its occupants in the collision? The front of the automobile is designed NOT to be rigid; it is built to collapse in sections, starting at the bumper, thus extending the collision distance (or time) as shown in Fig. 3.18a. The prolonged collapse reduces the deceleration force. The front of the car experiences severe damage, but the interior may be essentially undamaged with the consequence that its occupants may be bruised and shaken, but not seriously hurt. The amount of injury depends on additional safety features of the automobile, including seat belt systems and airbags which serve to protect the head and torso during a collision (Fig. 3.18b). Statistics indicate that these systems have been effective in reducing injury and death, but improper use of seatbelts and improper positioning of car seats for infants can produce the opposite result.

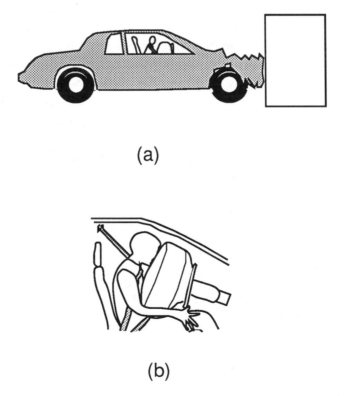

(a)

(b)

Figure 3.18. (a) An automobile is involved in a head-on collision and stops in a short distance. The deceleration distance can be about 1 m if the automobile is designed to collapse in the front end first. (b) The driver, who is wearing a harness seat belt, is rotated forward. An airbag inflates to cushion the driver's head and torso from collision with the steering wheel or dash.

Because of the hazards of uncontrolled automobile collisions, federal law requires a number of safety devices in automobiles. These include headrests, seat and shoulder belts (a three-point harness to prevent the person from being thrown from the automobile), energy-absorbing steering columns, penetration-resistant windshields, and side door beams to provide protection to the occupants during a side collision.

Information such as that given in Fig. 3.17 is used in the design of emergency escape methods from high-performance aircraft and in safety designs for commercial aircraft as well as for automobiles. For example, if a pilot is to be shot upward through an escape hatch, it is necessary to know the effects of acceleration in the seat-to-head direction. By knowing the limitations of the body, the accelerative force and its duration can be adjusted to minimize the probability of injury during emergency procedures.

A more familiar example of the use of the information in Fig. 3.17 is in the design of helmets for bicyclists, motorcycle riders, and for various sports such as baseball, football, hockey, and lacrosse. Each helmet is designed to reduce deceleration by crushing during impact. One criterion for bicycle helmets is the ability of the rider's head to withstand a 24 km/hr (15 mph) impact onto a rigid, flat surface as might happen if you fall when traveling at that speed. The helmet material must have the appropriate stiffness to compression so that the collapse of the helmet padding prolongs the deceleration and thus reduces the forces on the head. One must remember, however, that safety devices do not provide absolute protection.

3.11

PROBLEM

Estimate the force on the forehead in Fig. 3.18 if the mass of the head is 3 kg, its velocity is 15 m/s, and a padded dash is used instead of the air bag to stop the head in 0.02 s. [Answer: $F_{max} = 2.3 \times 10^3$ N]

3.4.4 Effects of Acceleration on Humans

Acceleration of the body produces a number of effects such as (1) an apparent increase or decrease in body weight, (2) changes in internal hydrostatic pressure, (3) distortion of the elastic tissues of the body, and

(4) the tendency of solids with different densities suspended in a liquid to separate. If the acceleration is sufficiently large, the body loses control because it does not have adequate muscle force to work against the large acceleration forces. Under certain conditions the blood may pool in various regions of the body; the location of the pooling depends upon the direction of acceleration. If a person is accelerated head first, the lack of blood flow to the brain can cause blackout and unconsciousness (see Chapter 8, *Physics of the Cardiovascular System*).

Astronauts in an orbiting satellite are in a condition of free fall or apparent weightlessness. Prior to man's first space flights, there were concerns about the physiological effects of weightlessness. Many of the effects predicted were based on changes observed in the body during extended periods of bed rest. We now have information about the effects on the body of extended time in space. Some physiological changes do take place; however, they have not been incapacitating or permanent.

Tissue can be distorted by acceleration and, if the forces are sufficiently large, tearing or rupture can take place. Laboratory information is sparse, but some experiments in huge centrifuges have shown that tissue can be stretched by accelerative forces until it tears. In some auto accidents, the aorta tears loose from the abdominal membrane leading to serious consequences if not death.

3.4.5 Oscillatory Motion

When walking, the legs (and arms) undergo a repetitive motion similar to that of a pendulum. Using this observation, we can estimate the speed of walking at a natural pace. We model the motion of the leg as a simple pendulum (ball at end of a string of length L) as illustrated in Fig. 3.19. The leg differs from the simple pendulum in that the mass of the leg is distributed nonuniformly, whereas the mass of the simple pendulum is concentrated at one point. To correct for this difference, we define the effective length of the leg, L_{eff}, as that length of a simple pendulum that would have the same period of oscillation as the complex shaped leg. (You might try to find this length by cutting out a model leg from heavy cardboard or other material and comparing its oscillation period with that of a simple pendulum whose length you can adjust so as to match periods.) For small oscillation amplitudes, the period of a simple pendulum is $T = 2\pi(L/g)^{1/2}$, where g is the acceleration of gravity. For a typical leg of a 2 m tall person, $L_{eff} = 0.2$ m and thus $T = 0.9$ s. (How does this agree with your natural walking pace? Remember, this is the time for one leg to return to the ground for the next step.) Since most of us have two legs, the time per step is T/2 = 0.45 s. If we assume that each step covers a distance of 0.9 m (about 3 ft) in 0.45 s, then our walking speed is

$$v = (0.9 \text{ m})/(0.45 \text{ s}) = 2 \text{ m/s} \ (7.2 \text{ km/hr or } 4.5 \text{ miles/hr})$$

Walking at a pace determined by the natural period of your leg uses the least amount of energy. Walking either faster or slower than this natural pace consumes more energy! Notice how much faster the step is for children and pets with their shorter legs.

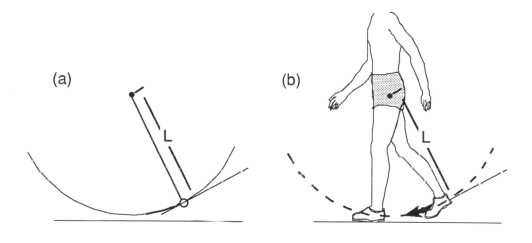

Figure 3.19. (a) A simple pendulum of length L undergoing small amplitude vibrations has a period $T = 2\pi(L/g)^{1/2}$. The quantity g is the acceleration of gravity. (b) The leg during walking also behaves like a pendulum. (After P. Davidovits, *Physics in Biology and Medicine*, Prentice-Hall, 1975, p. 47.)

Except for our bones, the organ systems in our body are composed mostly of water. Our organs are not securely fixed; they have flexible attachments to the skeleton. Each of our major organs has its own resonant frequency (or natural period) which depends on its mass and the elastic forces that act on it. Pain or discomfort occurs if a particular organ is vibrated vigorously at its resonant frequency (Fig. 3.20). Shock absorbers are devices to reduce or to dampen unwanted vibrational effects. Female athletes often use special bras to dampen the motion of their breasts

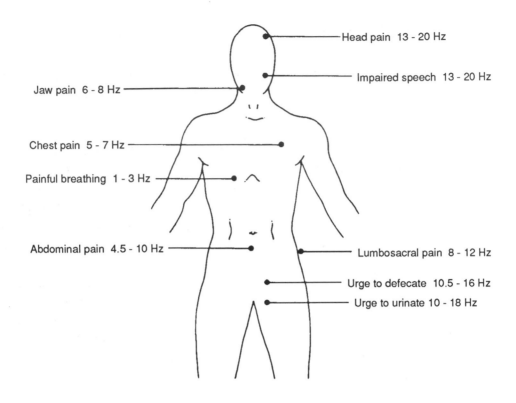

Figure 3.20. Pain symptoms of humans subjected to vibrations from 1 to 20 Hz. (Adapted from E. B. Magid, R. R. Coermann, and G. H. Ziegenruecker, "Human Tolerance to Whole Body Sinusoidal Vibration," *Aerospace Med.*, 31, 1960, p. 921.)

because they commonly jog at or near the natural frequency of the breast, which is about 2 Hz (1 Hz = 1 vibration/s), a period T = 0.5 s.

Excessive vibration often occurs in motor trucks and in some passenger aircraft. This results in fatigue and discomfort to the occupants, and may cause visual disturbances. The vibratory frequency of motorized vehicles, and of such subtle environmental systems as large fans used to distribute air in enclosed buildings, can be around 8 Hz or less, while those in aircraft are usually higher.

3.5 Physics of Teeth

As we grow into adults, our teeth undergo changes that usually do not concern us unless there is pain or expense. Toothaches and trips to the dentist cause concern, but most of the time our teeth play passive roles in our lives.

There are many applications of physics in our teeth and jaws—such as forces involved with biting, chewing, and erosion of teeth. In addition, *prosthetic* (replacement) devices such as bridges and crowns have to be biocompatible as well as have sufficient strength to function properly. Sometimes we inherit less than perfectly arranged teeth. We usually see an orthodontist who uses a variety of procedures applying force to reposition and straighten the teeth.

We consider first the physics of normal teeth, the forces involved in biting, and the force of the bite limited by the jaw (masseter) muscles. Next, we give simple examples of straightening and moving permanent teeth (*orthodontics*) and an example where the jaw is reshaped. Finally, we discuss a few prosthetic crowns and bridges.

3.5.1 Forces in Normal Teeth

Most of us wish we had teeth that were perfect. Figure 3.21 depicts the normal 32 permanent adult teeth and a cross section of a typical permanent molar tooth. It is obvious that different teeth have different functions. The *incisors* and *cuspids* (sometimes called eye teeth or canine teeth) have single cutting or biting edges. They have single roots; the roots for the cuspids located in the upper jaw are the longest. Behind the cuspids are the first and second *bicuspids*, followed by three molars, which usually have two or three roots. They are used for chewing or grinding food on the surface between the teeth (called the *occlusal surface*). Figure 3.22 shows a schematic view of the skull. The pivot for the jaw (*mandible*) is called the temporal-mandible joint (TMJ)—often a source of problems. The *masseter muscle* provides the main force for biting and chewing.

Scientists have measured the stress-strain behavior of the enamel and dentin components of teeth (see Chapter 4, *Physics of the Skeleton*, for definitions of stress-strain). A stress-strain curve for dentin is shown in Fig. 3.23. The maximum force than one can exert, measured at the first molar occlusal surface (1st bicuspid), is about 650 N. If the area of

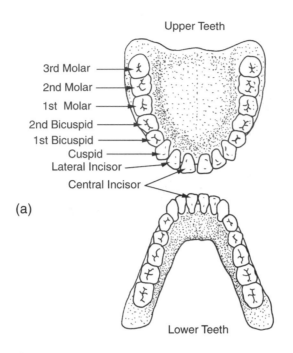

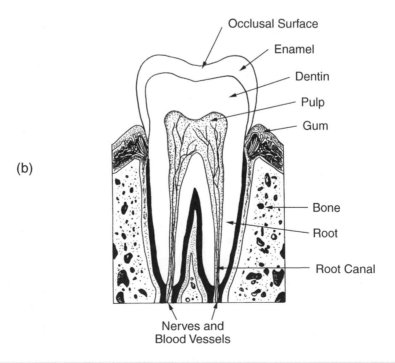

Figure 3.21. (a) The 32 normal permanent teeth of an adult. (b) Cross-section view of an adult molar tooth. (Images modified by Ken Ford, original image Copyright © 1994, TechPool Studios Corp. USA.)

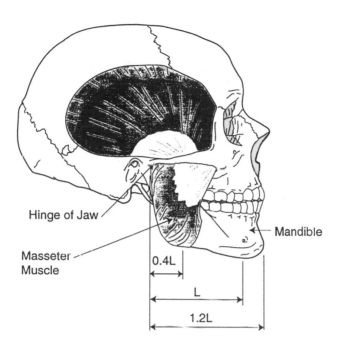

Figure 3.22. Schematic view of an adult skull showing some of the teeth and the masseter muscle that provides the closing and chewing action of the lower jaw (mandible). The dimensions are in units of L which is the distance of the first bicuspid from the hinge of the jaw. 0.4L is the approximate location of the masseter muscle from the hinge and 1.2L is the distance of the central incisor from the hinge. The value of L is typically about 6.5 cm for women and 8 cm for men. (Image modified by Ken Ford, original image Copyright © 1994, TechPool Studios Corp. USA.)

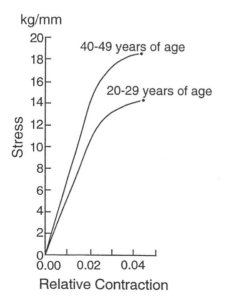

Figure 3.23. The stress-strain curve for wet dentin under compression for the premolar (bicuspids) teeth for adults in two different age groups. Young's modulus initially increases with age, but later it drops slightly. The enamel surface has a Young's modulus about five times greater than that for dentin. Note the stress scale is a factor of 10 larger if given in N/mm.^{2} [After H. Yamada, *Strength of Biological Materials*, F. H. Evans (ed.), Baltimore, Williams and Wilkins, 1970, p. 150 by permission.]

contact is about 10 mm^2, the force per unit area is then nearly 65 N/mm^2 (6.5×10^7 N/m^2 or kg/mm^2). The Hooke's law portion of the stress-strain curve in Fig. 3.23 in kg/mm^2 for dentin shows about a 0.01 (1%) fractional compression of the tooth. Considering that enamel is stronger than dentin by a factor of five, the biting force is well below that where failure of the tooth would occur.

If we accidentally bite into a hard cherry stone or kernel of popcorn, the area of contact may be as small as 1 mm^2; then the compressive stress is about 650 N/mm^2 (65 kg/mm^2). Under these conditions, the tooth would fail. Many of us have learned this fact experimentally. A tooth that has been weakened by fillings or decay might be broken when you bite a hard, small object.

The 650 N biting force is supplied by the masseter muscles. Going back to Fig. 3.4 we see that biting is a third class lever with the muscle close to the fulcrum of the jaw as shown in Fig. 3.22.

3.12

PROBLEM

From the dimensions of Fig. 3.22 and the force on the first bicuspid of 650 N, show that masseter force is 1625 N and the force on the central incisors is 540 N.

Because the molars are used for grinding food, they have large surface areas compared to the incisors, which act more like knives in the biting process. If the force from the masseter muscles were acting only on the central incisors and not on the molars, the net force would be less than 650 N by the ratio of L/1.2L or 540 N. This force is about equal to the weight of a small adult. You can imagine how effective the incisors would be when you visualize using a dull knife on an apple with a force about the same as the weight of a human.

Consider biting into an apple (see Fig. 3.24). The teeth behave like the knife shown in (a). When the incisors first make contact with the apple, the stress (force/area) is very large because of the large applied force (assume 200 N) and the small area of the edge of the incisor teeth (perhaps 1 mm^2). This applied force leads to a stress of 200 N/mm^2 (20×10^7 N/m^2), which is sufficiently large to rupture the apple (b) (and most other foods as well!). Once the apple's skin has been ruptured, then the front and back surfaces of the teeth make contact with the interior of the apple. The angle of the front incisors is about 60° as shown in (c) and (d) where the force of 200 N is still applied by the jaw on the front teeth. In the simplest approximation (d), the downward force is balanced by the two components of force F normal to the front and back surfaces of the incisors. These two forces can be large and push apart the two sides of the apple being bitten, causing the crack to spread.

3.13

PROBLEM

From the force diagram of biting in Fig. 3.24(d), estimate the forces normal to the two surfaces of the incisors.

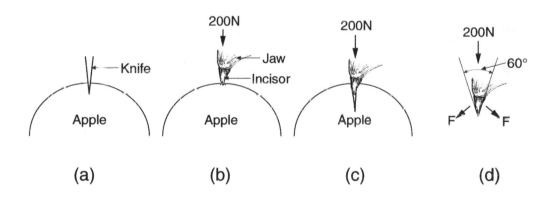

(a)   (b)   (c)   (d)

Figure 3.24. Schematic of the action of the biting behavior of an incisor tooth on an apple. (a) A knife cutting into an apple. (b) The incisor making contact with the apple and having sufficient force to cause the skin to rupture. (c) The incisor has penetrated the apple causing a crack to propagate. (d) The schematic behavior of the forces on the incisors.

Have you noticed that the permanent teeth of young adults appear very prominent in their jaws? This is particularly true of the front incisors. After 20 or 30 years this no longer seems to be the case. What has taken place is that the teeth wear; in some cases their length erodes as much as 0.1 mm/yr.

3.5.2 Some Simple Cases of the Physics in Orthodontics

Everyone has seen a child with its thumb in its mouth. It is part of growing up and nearly all children do this and eventually the thumb sucking ends. Excessive thumb sucking can change the shape of the mouth as it can move front teeth. Most often, the two central incisors are pushed out and spread apart, which can lead to a large overbite as shown in Fig. 3.25a.

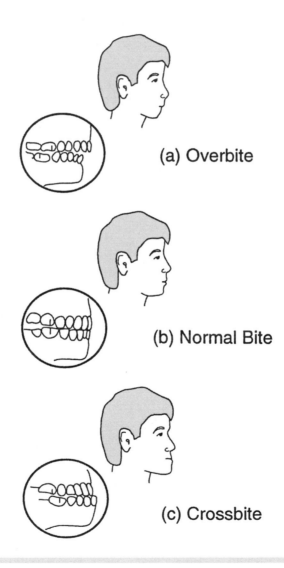

Figure 3.25. The location of the teeth in the upper jaw with respect to frontal teeth in the lower jaw leads to several conditions: (a) overbite, (b) normal bite, and (c) crossbite. (Image modified by Ken Ford, original image Copyright © 1994, TechPool Studios Corp. USA.)

How can those teeth be brought back to the desired location? One way is shown in Fig. 3.26a where a mechanical connection is made to the teeth that need to be moved and force is supplied by the external headgear. Depending upon the initial conditions of the teeth, other methods such as adding a rubber band to provide tension between the teeth (shown in Fig. 3.26b) may be all that is needed to move the teeth together. Sometimes a tooth needs to be moved a small amount; this can often be accomplished by appropriate spring wires as shown in Fig. 3.26c. It is surprising how small the force needs to be, in this case only about 1 N. However, we should remember that in the early years, the erupting teeth are guided by their surroundings: the jaw and the neighboring teeth.

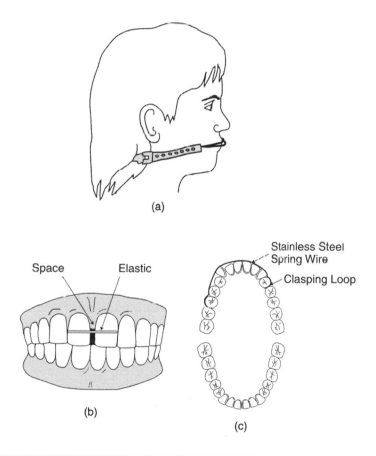

Figure 3.26. (a) Sometimes in orthodontic work the teeth are moved by an external headgear which supplies the force on the teeth. (b) For some teeth with too large a gap between them, the force of a rubber band is sufficient. (c) A simple brace (stainless spring steel) arrangement used to provide a small force (1–2 N) on a cuspid that needs to be moved into better alignment with the upper jaw. (Images modified by Ken Ford, original image Copyright © 1994, TechPool Studios Corp. USA.)

There are many orthodontic appliances and, to a large degree, they depend on the skill of the orthodontist. Figure 3.27 shows different methods to apply forces to move teeth. Figure 3.27a represents a fixed orthodontic apparatus. It has several features common to straightening and moving teeth, e.g., the banding and brackets are often used along with the arch wire to form the main support for other attachments to move teeth. Clever arrangements of the attachment bands, arch wire, and elastic bands can accomplish complicated movements of the teeth. Figure 3.27b depicts an adjustable, removable appliance designed to widen the jaws and straighten the front teeth. The adjustment moves about 0.8 mm per turn where each day one quarter of a turn is made. The total movement may be as much as one cm.

3.14

PROBLEM

Using Fig. 3.27a, give a descriptive discussion of the force directions and the desired changes for the teeth being moved or straightened.

Figure 3.28 shows two examples of moving teeth: (a) A spring under compression is used to widen the space for the middle tooth. (b) A spring under tension moves a tooth to close a gap. These springs supply a variable force for compression or expansion. Typically, the force will be about 1 N, which reduces as the tooth moves. Note that the springs are connected to the brackets attached to the teeth. The bracket on the tooth to be moved can slide guided by the arch wire; the other bracket is fixed to the arch wire.

3.5.3 Crowns, Bridges, and Implants

Despite our efforts to preserve our teeth, an accident can lead to broken teeth, or one or more teeth have decayed. We may be unfortunate and inherit genes that do not favor long-lasting teeth. Many people need

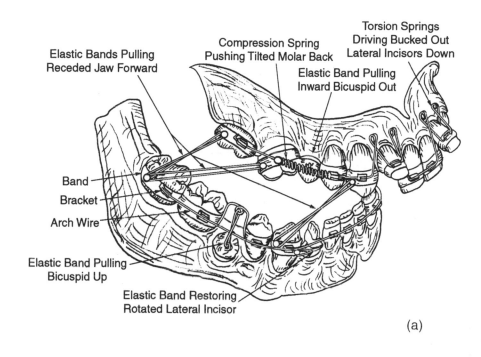

Torsion Springs
Driving Bucked Out
Lateral Incisors Down

Compression Spring
Pushing Tilted Molar Back

Elastic Bands Pulling
Receded Jaw Forward

Elastic Band Pulling
Inward Bicuspid Out

Band

Bracket

Arch Wire

Elastic Band Pulling
Bicuspid Up

Elastic Band Restoring
Rotated Lateral Incisor

(a)

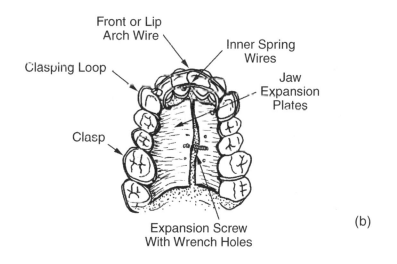

Front or Lip
Arch Wire

Inner Spring
Wires

Clasping Loop

Jaw
Expansion
Plates

Clasp

Expansion Screw
With Wrench Holes

(b)

Figure 3.27. Two schematic orthodontic arrangements. (a) An exaggerated case of a fixed orthodontic apparatus used to move and control teeth in the upper and lower right jaw (left side not shown). (b) An adjustable movable appliance used to widen the upper jaw while at the same time straightening the front teeth. This arrangement, when modified, can also be used to reduce the size of the jaw. (Adapted from S. Garfield, *Teeth, Teeth, Teeth*, New York, Simon & Schuster, 1969, p. 217)

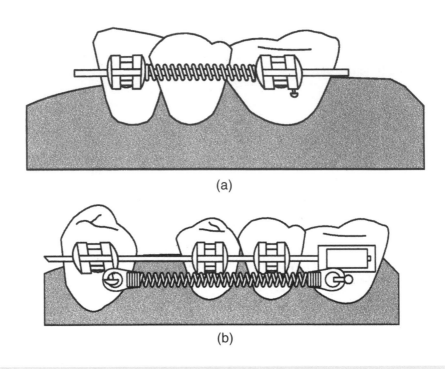

(a)

(b)

Figure 3.28. (a) A compressed spring arrangement is used to move an improperly aligned tooth to a different position. (b) A spring under tension supplies a force to move a tooth to fill a gap.

dental repair for damaged or missing teeth. The simplest repair is a simple filling. In many cases the filling does not significantly reduce the strength of the tooth; the repair may last a lifetime if properly done and given proper care.

Let us consider a more drastic case where the tooth has had extensive fillings and now is not structurally sound. How can one preserve the tooth and the function it provides? One approach is to *crown* the tooth, as shown in Fig. 3.29. This is a *prosthesis* and involves grinding away the damaged area of the tooth and replacing it with an artificial tooth. The shape of the crown is determined from molds made of the patient's mouth, ensuring a custom fit. The crown is often made of a strong gold alloy with a porcelain face, in a color matching the permanent teeth and cemented in place. The use of gold is not a new idea—the Etruscans, over 3000 years ago, made simple crowns. We now know that gold is inert chemically; it has a strength greater than the original teeth and can easily be cast in a mold. These repairs are attractive, functional, easy to keep clean, and long lasting.

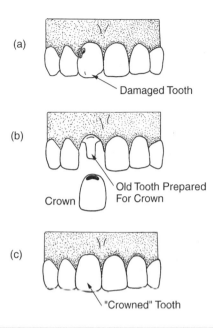

Figure 3.29. (a) A tooth damaged by decay. (b) The damaged tooth is prepared for a crown. An impression of the natural teeth Is used to prepare the crown replacement. (c) Shows the crown cemented in place. (Image modified by Ken Ford, original image Copyright © 1994, TechPool Studios Corp. USA.)

Suppose that the tooth has been damaged so much that it needs to be removed. You might be faced with the prospect of a "bridge" prosthesis. For this to work, there have to be teeth on both sides of the missing tooth for attaching the bridge. Figure 3.30 shows an example of a bridge which uses the adjacent teeth.

A bridge may fail when the material properties of the gold alloy are improperly used, such as an improper design with insufficient strength between the replacement tooth and its attachment to the neighboring teeth. If the cross-sectional area on both sides of the replacement tooth is insufficient, then the use of the bridge in chewing could cause the replacement tooth to flex eventually breaking the connections. Engineers call this a shearing force, which is another way of classifying the strength of materials.

What happens if a nearby tooth cannot be used for a support? Fig. 3.31 shows an implanted peg screwed directly into the jaw. The prosthetic tooth is then cemented to the peg. This type of prosthesis is more difficult to keep clean, but acceptable in many situations.

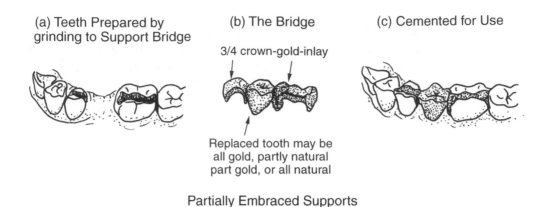

(a) Teeth Prepared by grinding to Support Bridge

(b) The Bridge

3/4 crown-gold-inlay

Replaced tooth may be all gold, partly natural part gold, or all natural

(c) Cemented for Use

Partially Embraced Supports

Figure 3.30. A simple bridge prosthesis. (a) The missing tooth and the teeth on either side prepared by grinding to support the bridge. (b) The bridge replacement and the supporting structures for either side of the missing tooth. A mold of the region of the missing tooth and of the teeth in the upper and lower jaw is needed to ensure the correct fit as shown in (c) for the bridge cemented into place. (Adapted from S. Garfield, *Teeth, Teeth, Teeth*, New York, Simon & Schuster, 1969, p. 257 by permission.)

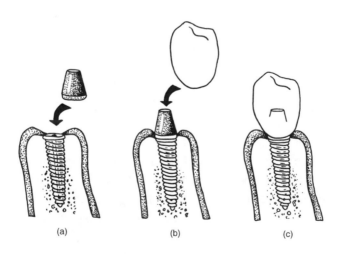

(a)   (b)   (c)

Figure 3.31. In a situation where there are no adjacent teeth to use for a bridge, sometimes an implant is used instead. (a) The implant is screwed into the jaw and the tissue and jaw are allowed to heal. (b) Later, the peg to hold the tooth is installed. (c) The finished tooth cemented in place. (Image modified by Ken Ford, original image Copyright © 1994, TechPool Studios Corp. USA.)

Of course, the materials for tooth repair need to be biocompatible. That is not a problem. Metals are often used to strengthen a body part (e.g., the hip transplant). The forces on prosthetic teeth go directly to the jaw like natural teeth. An implanted tooth is a very successful prosthesis.

4

Physics of the Skeleton

Anthropologists have been interested in bones for a long time. Bone can last for centuries and in some cases for millions of years. Because of its strength, bone has been used by humans for a wide variety of tools, weapons, and art objects. It provides the anthropologist with a means of tracing both the cultural and physical development of man.

Because of the importance of bone to the proper functioning of the body, a number of medical specialists are concerned with problems of bone. Two medical specialties, dentistry and orthopedic surgery, are completely devoted to this area. Other medical specialists who have considerable interest in bones are rheumatologists, M.D.s who specialize in problems of rheumatism and arthritis, chiropractors, and radiologists, who base many diagnostic decisions on x-ray images of bony structures.

Bones also are of interest to medical physicists and engineers. Perhaps bones appeal to physical scientists because they have engineering-type problems, e.g., static and dynamic loading forces that occur during standing, walking, running, lifting, and so forth. Nature has solved these problems extremely well by varying the shapes of the various bones of the skeleton (Fig. 4.1) and the types of bony tissue of which they are made. In adapting bone for different functions nature has done a better "design

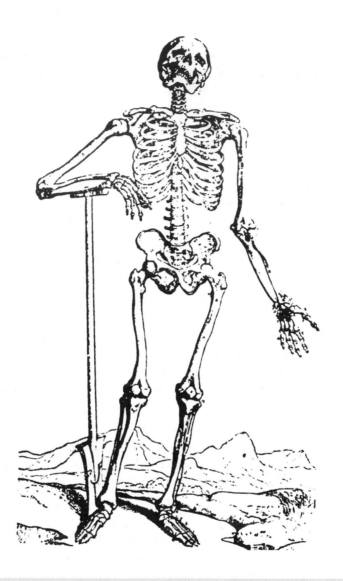

Figure 4.1. The skeleton of the body. (From A. Vesalius, *De Humani Corporais Fabrica*, Basle, 1543.)

job" than modern engineers are yet capable of doing. In fairness, it should be pointed out that nature has had many millions of years to refine its design, while man has only recently attempted to duplicate the functions and properties of bone for bone replacement. Bone has at least six functions in the body: (1) support, (2) locomotion, (3) protection of various organs, (4) storage of chemicals, (5) nourishment, and (6) sound transmission (in the middle ear).

The support function of bone is most obvious in the legs. The body's muscles are attached to the bones through tendons and ligaments and the system of bones plus muscles supports the body. In old age and in certain diseases some of this support structure deteriorates. If we lived in the sea where we would be almost "weightless" due to the buoyancy of the water, our need for a bony skeleton would be greatly reduced. Sharks do not have any bones; their skeleton is made of cartilage.

Bone joints permit movement of one bone with respect to another. These hinges, or articulations, are very important for walking as well as for many of the other motions of the body. We can manage with some loss of joint movement, but the destruction of joints by arthritis can seriously limit locomotion as well as use of our hands.

Protection of delicate body parts is an important function of some bones. The skull, which protects the brain and several of the most important sensory organs (eyes and ears), is an extremely strong container. The ribs form a protective cage for the heart, lungs, and liver. (The ribs and muscles of the chest also act as a bellows-like structure, which upon expansion and contraction allows the inhalation and expiration of air.) In addition to its support role, the spinal column serves as a flexible shield for the spinal cord.

Bones act as a chemical "bank" for storing elements for future use by the body. The body can withdraw these chemicals as needed. For example, a minimum level of calcium is needed in the blood; if the level falls too low, a "calcium sensor" causes the parathyroid glands to release more parathyroid hormone into the blood, and this in turn causes the bones to release the needed calcium.

The teeth are specialized bones that can cut food (incisors), tear food (canines), and grind food (molars) and thus serve in providing nourishment for the body (see also Section 3.5). In man they come in two sets—deciduous (baby) teeth and permanent teeth (a third set is sometimes obtained from a dentist).

The smallest bones of the body are the ossicles in the middle ear. These three small bones act as levers and provide an impedance matching system for converting sound vibrations in air to sound vibrations in the fluid in the cochlea (see Chapter 11, *Physics of the Ear and Hearing*). They are the only bones that attain full adult size before birth!

It is sometimes thought that bone is a rather dead or inert part of the body and that once it has reached adult size, it remains the same until death or some other calamity (such as a skiing accident) strikes. Actually, bone is a living tissue and has a blood supply as well as nerves. Most of the bone tissue is inert, but distributed through it are the osteocytes, cells that maintain the bone in a healthy condition. Cells make up about 2% of the volume of bone. If these cells die (e.g., due to a poor blood supply), the bone dies and it loses some of its strength. A serious hip problem is caused by a condition called aseptic necrosis in which the bone cells in the hip die due to lack of blood. The hip joint usually fails to function properly and sometimes has to be replaced with an artificial joint.

Since bone is a living tissue, it undergoes change throughout life. A continuous process of destroying old bone and building new bone, called bone remodeling, is performed by specialized bone cells. Osteoclasts destroy the bone, and osteoblasts build it. Compared to many body processes, bone remodeling is slow work. We have the equivalent of a new skeleton about every seven years; each day the osteoclasts destroy bone containing about 0.5 g of calcium (the bones have about 1000 g of calcium), and the osteoblasts build new bone using about the same amount of calcium. While the body is young and growing, the osteoblasts do more than the osteoclasts; but after the body is 35 to 40 years old, the activity of the osteoclasts is greater than that of the osteoblasts, resulting in a gradual decrease in bone mass that continues until death.* This decrease is usually faster in women than in men and leads to a serious problem of weak bones in older women. This condition, called osteoporosis (literally, porous bones), results in spontaneous fractures, especially in the spine and hips. In Section 4.4 we discuss how this disease can be diagnosed and studied through the use of a physical measurement.

4.1   The Composition of Bone

The detailed chemical composition of bone is given in Table 4.1. Note the relatively large percentage of calcium (Ca) in bone. Since calcium has a much heavier nucleus than most elements of the body, it absorbs x-rays much better than the surrounding soft tissue. This is the reason x-rays show bones so well (Fig. 4.2).

Bone consists of two quite different materials plus water: collagen, the major organic fraction, which is about 40% of the weight of solid bone and 60% of its volume, and bone mineral, the so-called "inorganic" component of bone, which is about 60% of the weight of the bone and

---

*As in other aspects of life, destruction is easier than construction. One osteoclast can destroy bone 100 times faster than one osteoblast can build new bone.

Table 4.1. Composition of Compact Bone[*]

Element	Compact Bone, Femur (%)
H	3.4
C	15.5
N	4.0
O	44.0
Mg	0.2
P	10.2
S	0.3
Ca	22.2
Miscellaneous	0.2

[*]Adapted from H. Q. Woodard, *Health Physics*, 8, 516 (1962), by permission of the Health Physics Society and the author.

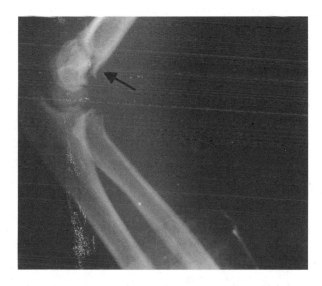

Figure 4.2. An x-ray of the upper and lower arm showing a break just above the elbow (arrow). The lower part of the arm was broken earlier and is covered with a cast.

40% of its volume. Either of these components may be removed from bone, and in each case the remainder, composed of only collagen or bone mineral, will look like the original bone. The collagen remainder is quite flexible, somewhat like a chunk of rubber, and can even be bent into a loop (Fig. 4.3). While it has a fair amount of tensile strength, it bends easily if it is compressed. When the collagen is removed from the bone, the bone mineral remainder is very fragile and can be crushed with the fingers! A simple way to remove the collagen is to put the bone in a furnace and "ash" it. Cremation is the ashing of the whole body; the bone mineral is the matter that is put in an urn.

Collagen is apparently produced by the osteoblastic cells; mineral is then formed on the collagen to produce bone. Bone collagen is not the same as the collagen found in many other parts of the body such as the skin. Its structure corresponds to the crucial dimensions of the crystals of bone mineral, and it forms a template onto which the bone mineral crystals fit snugly.

Bone mineral is believed to be made up of calcium hydroxyapatite $Ca_{10}(PO_4)_6(OH)_2$. Similar crystals exist in nature; fluorapatite, a common rock, differs from calcium hydroxyapatite in that fluorine takes the place of the OH. Fluorine in drinking water may prevent caries, or cavities in the teeth, by turning microscopic areas of the teeth into the rock fluorapatite, which is more stable than bone mineral.

Studies using x-ray scattering indicate that the bone mineral crystals are rod shaped with diameters of 2 to 7 nm and lengths of from 5

Figure 4.3. If the bone mineral in a bone is dissolved with a 5% acetic acid solution, the remaining collagen is quite flexible. Here Nancy Clark easily bends an adult tibia (shin bone) that has been demineralized by this method.

to 10 nm (1 nm = 10^{-9} m). (The nanometer (nm) is a convenient unit for measuring atomic dimensions, since many atoms have diameters of about 0.1 nm.) Because of the small size of the crystals, bone mineral has a very large surface area. In a typical adult, it has a surface area of over 4×10^5 m^2 (~100 acres)—roughly the area of 12 square city blocks! Around each crystal is a layer of water containing in solution many chemicals needed by the body. The large area of exposed bone mineral crystal permits the bones to interact rapidly with chemicals in the blood and other body fluids. Within a few minutes after a small quantity of radioactive fluorine (^{18}F) is injected into a patient, it will be distributed throughout the bones of his body. Bone tumors not yet visible on an x-ray can be identified by this method. Bone in a bone tumor is being destroyed somewhat like a brick house being torn down a brick at a time. When the radioactive fluorine atoms come in contact with this partially destroyed bone, they find many places they can fit in more so than in normal bone. The increased radiation from the tumor area signals the possibility of a bone tumor.

4.2 The Strength of Bones

If a mechanical engineer were confronted with the problem of designing the skeleton, the engineer would, of course, need to examine the functions of the different bones since their functions would determine their shapes, their internal construction, and the type of material to be used. We have discussed a number of the more obvious functions of bones in the body. Now let us look at how bones have developed to meet our needs. If you sorted all of the approximately 200 bones of the body into various piles according to their shapes, you might come up with five piles: a small pile of flat, plate-like bones such as the shoulder blade (scapula) and some of the bones of the skull; a second pile of long, hollow bones such as those found in the arms, legs, and fingers; a third pile of more or less cylindrical bones from the spine (vertebrae); a fourth pile of irregular bones such as from the wrist and ankle; and a fifth pile of bones such as the ribs that do not belong in any of the other piles.

If you were to saw some of the bones you would find that they are composed of one or a combination of two quite different types of bone: solid (or compact) bone and spongy (or cancellate) bone made up of thin threadlike trabeculae—trabecular bone. Fig. 4.4a shows these two types of bone in an adult femur sawed along its long axis. Trabecular bone is

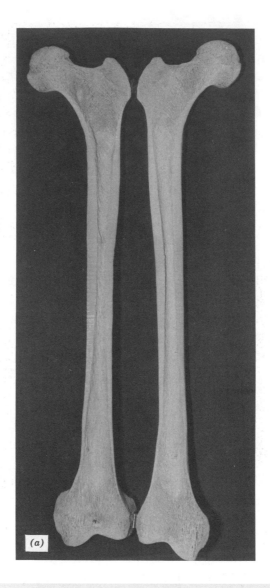

Figure 4.4. Cross sections of (a) an adult femur (thigh bone), (b) a normal vertebra cut vertically, and (c) an osteoporotic vertebra (from an 80-year-old woman) cut vertically. Note the arrangement of compact and trabecular bone. (Vertebra figures courtesy of B. L. Riggs, M.D., Mayo Clinic, Rochester, MN.)

predominately found in the ends of the long bones, while most of the compact bone is in the central shaft. Fig. 4.4b shows a cross section of a normal vertebra; note that it is almost entirely trabecular bone with the exception of thin plates of compact bone on the surfaces. Trabecular bone is considerably weaker than compact bone due to the reduced amount of

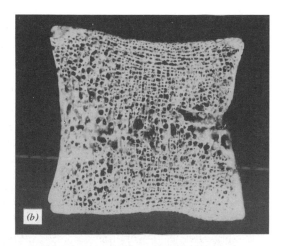

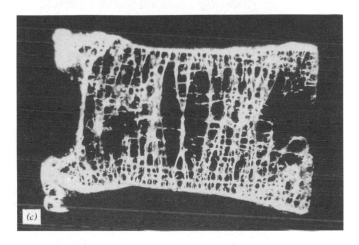

bone in a given volume. Osteoporotic bone (Fig. 4.4c) is even weaker. On a microscopic level the bone tissue in a trabecular bone is the same as that in compact bone.

A study of the construction of the femur illustrates how well it is designed for its job. Stress (force per unit area) in a bone can be analyzed the same way as stress in a beam in a building. Fig. 4.5a shows a horizontal beam supported at the ends with a downward force in the middle. The stresses inside the beam (shown by arrows) are pulling it apart at the bottom (tension) and pushing it together at the top (compression). There is relatively little stress of either type in the center of the beam. For this reason it is common to use an I beam, which has a thick top and bottom joined with a thin web, as a support beam in a building (Fig. 4.5b). When the force may come from any direction, a hollow cylinder is used to get the maximum strength with a minimum

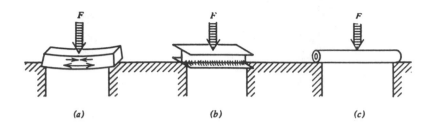

Figure 4.5. Various types of beams subjected to force F. (a) In a simple rectangular beam the greatest stresses are near the top and bottom. There is little stress in the middle of the beam. (b) Because the stress in the middle is small, a beam that has less material there—an I beam—can be used. (c) A tubular beam can be thought of as a rotated I beam with the center web removed. It is used when the force may come from any direction.

amount of material (Fig. 4.5c). It is almost as strong as a solid cylinder of the same diameter. Since the forces on the femur may come from any direction, the hollow cylinder structure of the bone is well suited for support. If you push on one end of a hollow cylinder such as a soda straw, it will tend to buckle near the middle rather than at either end. Extra thickness at the middle would strengthen it. The compact bone of the shaft of the femur is thickest in the center and thinnest at the ends (Fig. 4.4a); note again the high quality of the design.

The trabecular patterns at the ends of the femur are also optimized for the forces to which the bone is subjected. Fig. 4.6a shows schematically the lines of tension and compression in the head and neck of the femur due to the weight on the head. Fig. 4.6b shows a cross section of this part of the femur; notice that the trabeculae lie along the lines of force shown in Fig. 4.6a. Similarly, in the lower (distal) end of the femur the forces are nearly vertical, as are the trabeculae. There is some cross-banding to reinforce the trabeculae.

There are at least two advantages of trabecular bone over compact bone. Where a bone is subjected primarily to compressive forces, such as at the ends of the bones and in the spine, trabecular bone gives the strength necessary with less material than compact bone. Also, because the trabeculae are relatively flexible, trabecular bone in the ends of long bones can absorb more energy when large forces are involved such as in walking, running, and jumping. On the other hand, trabecular bone cannot withstand very well the bending stresses that occur mostly in the central portions of long bones.

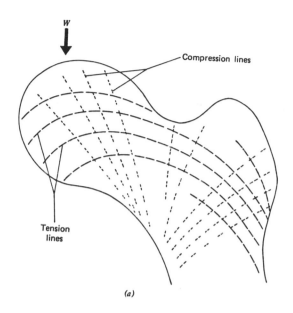

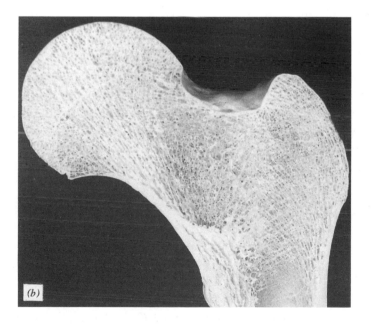

Figure 4.6. The head and neck of the femur. (a) The lines of compression and tension due to the weight W of the body. (b) A cross section showing the normal trabecular patterns. Note that they follow the compression and tension lines.

Now let us consider some of the mechanical properties of bone, a composite material analogous to fiberglass. As described in Section 4.1, bone is composed of small, hard, bone mineral crystals attached to a soft, flexible collagen matrix. These components have vastly different mechanical properties that also differ from those of bone. The exact nature of the interplay of these two components in producing the remarkable mechanical properties of bone is unknown. Nevertheless, the combination provides a material that is as strong as granite in compression and 25 times stronger than granite under tension.

We can make some standard physics and engineering measurements on a piece of compact bone, such as determining its density (or specific gravity); how much it lengthens or compresses under a given force (Young's modulus of elasticity); and how much force is needed to break it by compression, tension, and twisting. We may also determine how its strength depends on the time over which the force is applied and how much elastic energy is stored in it just before it breaks. The density of compact bone is surprisingly constant throughout life at about 1.9×10^3 kg/m^3 (or 1.9 times as dense as water). In old age the bone becomes more porous and disappears from the inside, or endosteal, surface of solid bone. The density of the remaining compact bone is still normal; it is reduced in strength because it is thinner, not because it is less dense. The physical quantity *bone density* is often confused with *bone mass*. An x-ray of a bone gives an idea of the mass of the bone, not of its density. The confusion is partly due to the use of density in connection with the optical density of an x-ray image. (In Section 4.4 we discuss instrumentation for measuring bone mass and bone density in patients.)

All materials change in length when placed under tension or compression. When a sample of fresh bone is placed in a special instrument for measuring the elongation under tension, a curve similar to that in Fig. 4.7 is obtained. The strain, ($\Delta L/L$), increases linearly at first, indicating that it is proportional to the stress (F/A)—Hooke's Law. As the force increases, the length increases more rapidly, and the bone breaks at a stress of about 1.2×10^8 N/m^2 (~17,000 lb/in.^{2}). The ratio of stress to strain in the initial linear portion is Young's modulus, Y. That is

$$Y = \frac{LF}{A\Delta L} \qquad (4.1)$$

Young's moduli for bone and a few common structural materials are given in Table 4.2. It is usually of more interest to calculate the change in length, L, for a given force, F. Equation 4.1 can be rewritten as

$$\Delta L = \frac{LF}{AY} \qquad (4.2)$$

Equations 4.1 and 4.2 are valid for both tension and compression (see Problem 4.1).

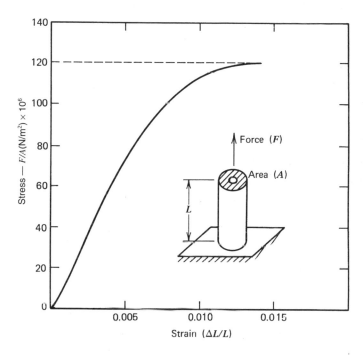

Figure 4.7. When a piece of bone is placed under increasing tension, its strain $\Delta L/L$ increases linearly at first (Hooke's Law) and then more rapidly just before it breaks in two at about 120×10^6 N/m^2 (~17,000 lb/in.^{2}).

Table 4.2. Strengths of Bone and Other Common Materials

Material	Compressive Breaking Stress ($\times 10^8$ N/m^2)	Tensile Breaking Stress ($\times 10^8$ N/m^2)	Young's Modulus of Elasticity ($\times 10^8$ N/m^2)
Hard Steel	5.520	8.270	2070.
Rubber	—	0.021	0.01
Granite	1.450	0.048	517.
Concrete	0.210	0.021	165.
Oak	0.590	1.170	110.
Porcelain	5.520	0.055	—
Compact Bone	1.700	1.200	179.
Trabecular Bone	0.022	—	0.76

The ability of the bones to support the body's weight without breaking is crucial to human mobility and well-being. Of course, they support not only weight but also other forces. In bending over to pick up a heavy object we may develop large forces in the lower spine (Chapter 3, Section 3.3.5). This explains why crushed vertebrae of the lower (lumbar) spine are common (Fig. 4.8). Large forces are also produced in such activities as running and jumping. In running, the force on the hip bone when the heel strikes the ground may be four times the body's weight. Even in normal walking the forces on the hip are about twice the body's weight.

What is the built-in safety factor in the bones that support the body's weight? Engineers like to over-design a support structure so that it can withstand forces of about 10 times the maximum expected forces. How well does the femur meet this requirement? Healthy compact bone is able to withstand a compressive stress of about 1.7×10^8 N/m^2 (~25,000 lb/in.^{2}) before it fractures (Table 4.2). The mid-shaft of the femur has a cross-sectional area of about 3.3×10^{-4} m^2 (0.5 in.^{2}); it would support a force of about 5.7×10^4 N (~12,000 lb, or 6 tons)! The

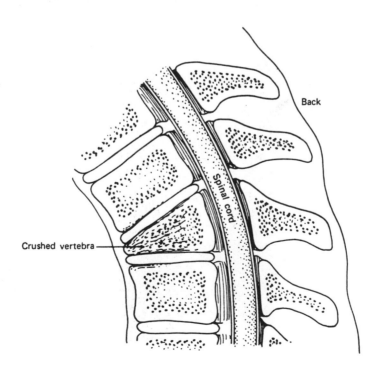

Figure 4.8. Diagram of a crushed lumbar vertebra. The resulting curvature of the spine produces a hunch-backed appearance.

cross-sectional area of the shin bone (tibia) is not as great, but the safety margin is satisfactory for most activities except downhill skiing.

The bones do not normally break due to compression; they usually break due to shear (Fig. 4.9a and b) or under tension (Fig. 4.9c). A common cause of shear is catching the foot and then twisting the leg while falling. A shear fracture often results in a spiral break (Fig. 4.9b) in which the bone is apt to puncture the skin. Such a compound fracture is more apt to become infected than a fracture in which the bone is not exposed—a simple fracture.

The bones are not as strong under tension as they are in compression; a tension stress of about 1.2×10^8 N/m^2 ($\sim$17,000 lb/in.^{2}) will cause a bone to break (Fig. 4.7). However, compact bone is stronger under tension than many common materials (Table 4.2).

Let us consider the forces exerted on a bone during a fall. From Newton's second law, the force exerted during a collision or a fall is equal to the rate of change of momentum, which is simply the momentum of the body divided by the duration of impact (Chapter 3, *Muscle and Forces*). Therefore, the shorter the duration of impact, the greater the force. To reduce the force and thereby reduce the likelihood of fracture, it is necessary to increase the impact time. In both falling down

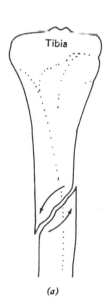
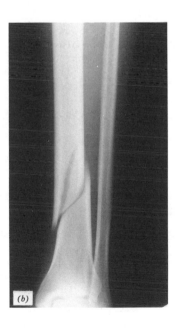
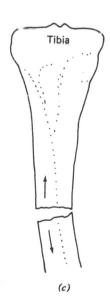

(a)   *(b)*   *(c)*

Figure 4.9. Fractures of the tibia. (a) A schematic of a spiral fracture caused by shear (twisting). (b) An x-ray of a spiral fracture caused by shear. (c) A schematic of a tension fracture in the tibia.

and jumping from an elevation, the impact time can be increased significantly by simply rolling with the fall or jump, thereby spreading the change in momentum of the body over a longer time. A good example of rolling with the impact is the manner in which a parachutist is trained to land: the ankles and knees bend upon impact and the body turns to one side and falls first on the leg, then on the hip, and then on the side of the chest. In a stiff-legged landing, the force generated would be about 1.42×10^5 N (32,000 lb), which means that each tibia, which is about 3.3×10^{-4} m^2 in area at the ankle, would bear a stress of about 2.1×10^8 N/m^2 (31,000 lb/in.^{2}). This value exceeds the maximum compression strength of bone by about 30%.

Bone, however, can withstand a large force for a short period without breaking, while the same force over a long period will fracture it. That is, the short-term force developed when you fall or jump, while possibly exceeding the maximum compressive strength of bone, is not as dangerous as the same force applied over a longer period of time. This property is called *viscoelasticity*.

When a bone is fractured, the body can repair it if the fracture region is immobilized. Even in an elderly woman with osteoporosis the healing process is effective. However, the long period of bed confinement necessary for a fractured hip to heal is very debilitating, and it is important to get the patient on her feet as soon as possible. Metal prosthetic hip joints, pins, nails, and so forth, are often used to repair such damaged bones (Fig. 4.10).

While the physics of the growth and repair of bone is not well understood, there is good evidence that local electrical fields may play a role. When bone is bent, it generates an electrical charge on its surfaces. It has been suggested that this phenomenon (piezoelectricity) may be the physical stimulus for bone growth and repair. Experiments with animal bone fractures have shown that bone heals faster if an electrical potential is applied across the break. Producing an electric current across a break requires electrical leads to enter the body, which would be a source of infection. It is possible to produce currents in and near the break using pulsed magnetic fields. This technique appears to be successful. However, a double blind study of the technique suggests that it works as well with or without the magnetic pulses! The improvement in healing might be the result of the additional restraint of the limb by the pulsing equipment for many hours per day.

4.1

PROBLEM

Assume a leg can be represented as a 1.2 m shaft of bone with an average cross-sectional area of 3 cm^2 (3 × 10^{-4} m^2). What is an amount of shortening when a person of body mass M = 70 kg supports all of the body weight on this leg?
[Answer: ΔL = 1.5 × 10^{-4} m = 0.15 mm]

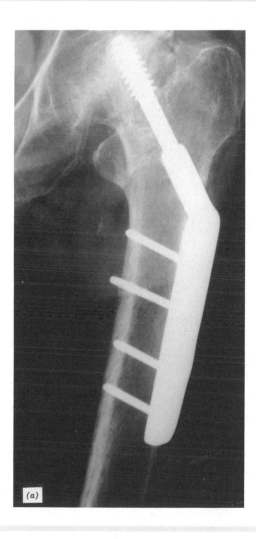

Figure 4.10. Hip prostheses. (a) A weak hip joint can be reinforced by a metal support fastened to the femur. In this instance the femoral neck has been made shorter to reduce the stress.

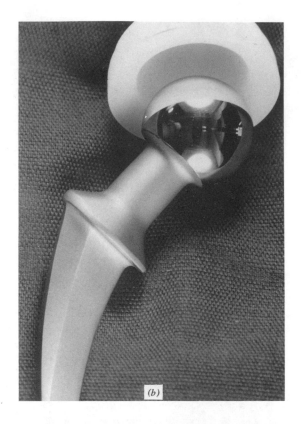

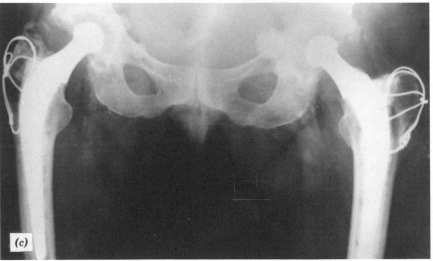

Figure 4.10 (continued.) (b) The entire hip joint can be replaced with man-made material. (c) An x-ray of a double hip replacement using prostheses similar to the prosthesis shown in b.

4.2

PROBLEM

Using the information given in Fig. 4.7,

(a) Calculate the maximum tension a bone with a cross-sectional area of 4 cm^2 could withstand just prior to fracture. [Answer: 4.8×10^4 N/m^2]

(b) Determine the elongation of a bone whose initial length is 0.35 m under the maximum tension you found in (a).
[Answer: 5×10^{-3} m, or 5mm]

(c) Calculate the stress on this bone if a tension force of 10^4 N were applied to it. How much would this bone elongate?
[Answer: 2.5×10^7 N/m^2, 5×10^{-4}m, or 0.5 mm]

4.3 Lubrication of Bone Joints

Those of us who do not suffer from arthritis take our well-functioning bone joints for granted. Many people are not as fortunate. An analysis of 1000 autopsy reports revealed that over two-thirds of the cadavers had a joint problem in the knee and that about one-third had a similar problem in the hip. There are two major diseases that affect the joints—rheumatoid arthritis, which results in over-production of synovial fluid in the joint and commonly causes swollen joints, and osteoarthrosis, a disease of the joint itself.

The lubrication of bone joints is not understood in detail, but the essential features are agreed upon. The main components of a joint are shown in Fig. 4.11. The synovial membrane encases the joint and retains the lubricating synovial fluid. The surfaces of the joint are articular cartilage, a smooth, somewhat rubbery material that is attached to the solid bone. A disease that involves the synovial fluid, such as rheumatoid arthritis, quickly affects the joint itself.

The surface of the articular cartilage is not as smooth as that of a good man-made bearing. It has been suggested that its roughness plays a useful role in joint lubrication by trapping some of the synovial fluid. It has also been suggested that because of the porous nature of the articular cartilage, other lubricating material is squeezed into the joint when it is under its greatest stress—when it needs lubrication the most. One theory is that pressure causes lubricating "threads" to squeeze out of the cartilage into the joint; one end of each lubricating thread remains in the cartilage, and as the pressure is reduced the threads pull back into

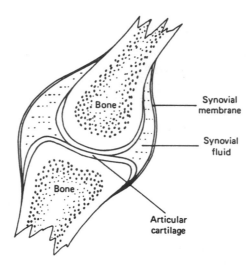

Figure 4.11. The main components of a joint.

their holes, somewhat like nightcrawler worms when you try to catch them. This *boosted lubrication* is a technique engineers have not yet been able to adapt to industry.

The lubricating properties of a fluid depend on its viscosity; thin oil is less viscous and a better lubricant than thick oil. The viscosity of synovial fluid decreases under the large shear stresses found in the joint. The good lubricating properties of synovial fluid are thought to be due to the presence of hyaluronic acid and mucopolysaccharides (molecular weight, ~500,000) that deform under load.

The coefficient of friction of bone joints is difficult to measure under the usual laboratory conditions. Little, Freeman, and Swanson[1] described the arrangement shown in Fig. 4.12. A normal hip joint from a fresh cadaver was mounted upside down with heavy weights pressing the head of the femur into its socket. The weight on the joint could be varied to study the effects of different loads. The whole unit acted like a pendulum with the joint serving as the pivot. From the rate of decrease of the amplitude with time, the coefficient of friction was calculated. The coefficient of friction was found to be independent of the load from 89 to 890 N (20 to 200 lb) and independent of the magnitude

---

[1]In Wright, V. (Ed.), *Lubrication and Wear in Joints*, Proceedings of a Symposium organized by the Biological Engineering Society and held at The General Infirmary, Leeds, on April 17, 1969, Lippincott, Philadelphia, 1969.

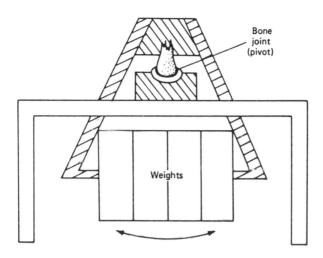

Figure 4.12. Arrangement for determining the coefficient of friction of a joint. The joint is used as the pivot in a pendulum and the decrease in amplitude of the oscillations with time is measured.

of the oscillations. It was concluded that fat in the cartilage helps to reduce the coefficient of friction. For all healthy joints studied, the coefficient of friction was found to be less than 0.01, much less than that of a steel blade on ice, which is 0.03. (A coefficient of friction of 0.01 means that if there is a 100 lb force on a joint, only 1 lb of force is needed to move it.) When the synovial fluid was removed, the coefficient of friction increased considerably.

4.4 Measurement of Bone Mineral in the Body

Bone is one of the most difficult organs to study. With the exception of the teeth, the bones are relatively inaccessible. In this section we describe a physical system for accurately measuring the bones *in vivo* (in the living body). There are many other physical techniques for studying bone, but most are used on excised bone samples (*in vitro* studies). Bone disease is one of the most common problems of the elderly. For example, each year about 200,000 women in the United States break a hip. Most of these women are elderly and have osteoporosis. In the 1960s, osteoporosis was difficult to detect until a patient appeared with a broken hip or a crushed vertebra. At that time it was too late to use preventive therapy.

The strength of bone depends to a large extent on the mass of bone mineral present, and the most striking feature in osteoporosis is the lower than normal bone mineral mass. Thus a simple technique to measure bone mineral mass *in vivo* with good accuracy and precision (reproducibility) was sought. It was hoped that such a technique could be used to diagnose osteoporosis before a fracture occurred and also to evaluate various types of therapy for osteoporosis. Since bone mineral mass decreases very slowly, 1 to 2% per year, a very precise technique was needed to show changes.

The idea of using an x-ray image to measure the amount of bone mineral present is an old one; it was first tried in 1901! The major problems of using an ordinary x-ray (Fig. 4.13) are (1) the x-ray beam has many different energies, and the absorption of x-rays by calcium varies rapidly with energy in this range of energies; (2) the relatively large beam contains much scattered radiation when it reaches the film; and (3) the film is a poor detector for making quantitative measurements since it is nonlinear with respect to both the amount and the energy of

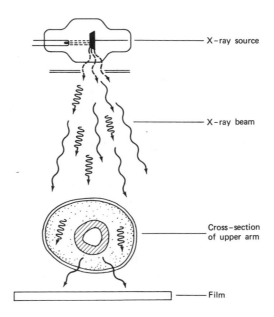

Figure 4.13. Conventional x-rays are not useful for quantitative measurement of bone mineral because the beam is heterogeneous, the scatter in the image is unknown, and film is not a reproducible detector.

the x-rays. Developing the film can introduce additional variations. The net result of these problems is that a large change in bone mineral mass (30 to 50%) must occur between the taking of two x-rays of the same patient before a radiologist can be sure that there has been a change. Each of the problems can be reduced by special methods, but the determination of bone mineral mass by this technique (x-ray film densitometry) has been limited to only a few laboratories in the world.

An improved technique based on the same physical principles was invented by one of the authors (JRC) starting about 1960. The basic components used in this technique, called photon absorptiometry, are shown in Fig. 4.14. The three problems with the x-ray technique were largely eliminated by using (1) a monoenergetic x-ray or gamma ray source, (2) a narrow beam to minimize scatter, and (3) a scintillation detector that detects all photons and permits them to be sorted and

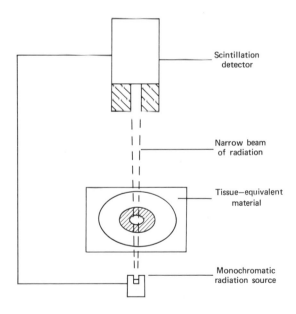

Scintillation detector

Narrow beam of radiation

Tissue—equivalent material

Monochromatic radiation source

Figure 4.14. The basic components used in photon absorptiometry. A radionuclide that emits essentially only one energy, such as Iodine 125 (27 keV) or Americium 241 (60 keV), serves as the radiation source; the limb is embedded in a uniform thickness of tissue-equivalent material; and the transmitted fraction of the narrow beam is detected by a scintillation detector.

counted individually. The determination of the bone mineral mass is further simplified by immersing the bone to be measured in a uniform thickness of soft tissue (or its x-ray equivalent, e.g., water).

Fig. 4.15 shows a graph of the logarithm of the transmitted intensity, I, of the beam (log I) as it scans across a bone immersed in a uniform thickness of "tissue." The intensity before the beam enters the bone is called I_0^*. The bone mineral mass (BM) at any point in the beam is proportional to log (I_0^*/I) and is given by

$$BM(g/cm^2) = k \log (I_0^*/I) \qquad (4.3)$$

where k is a constant that can be determined experimentally. This calculation is done electronically for all points in the beam, and the results are integrated to give the bone mineral mass of the slice of bone in grams per centimeter length of bone. A modern clinical bone scanner that uses the photon absorption technique is shown in Fig. 4.16. The unit has a reproducibility of 1 to 2% when used by a trained operator.

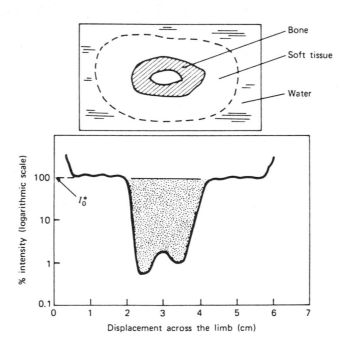

Figure 4.15. A graph of the transmitted intensity of the beam as it traverses the bone during photon absorptiometry. The intensity is plotted on a logarithmic scale. The shaded area is proportional to the bone mineral mass per cm length of bone.

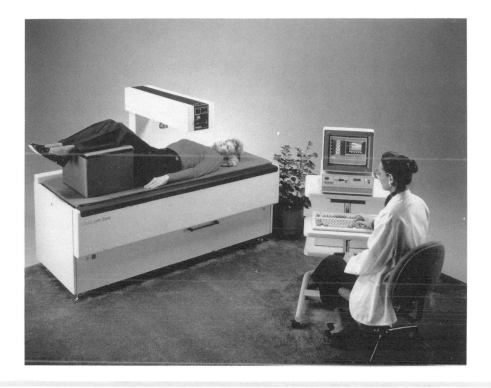

Figure 4.16. Dual x-ray energy bone densitometer used to measure the lumbar spine. A 2 to 5 minute examination time is needed. This technique provides high precision (1%) with a low entrance skin dose of 10 μGy (1 millirad). Photograph provided by Lunar Corporation, Madison, Wisconsin. Used by permission.

This equipment does not work where there is not a uniform tissue thickness. Thus it is not useful for measuring the bone mineral of the spine. It is, however, possible to use two different energies to measure bone mineral of the spine. This is called *dual energy absorptiometry*. Originally, it was done with a radioactive source such as Gd-153, but it is now possible to use a special x-ray generator with two different energies.

5

Pressure in the Body

Pressure is a very common phenomenon in our lives. The weatherman tells us the atmospheric pressure, the service station attendant checks the pressure in our tires, and the doctor measures our blood pressure as part of a physical examination. Pressure is used for gases or liquids. In solids, the quantity force per unit area is referred to as *stress*.

Pressure (P) is defined as the force per unit area (P = F/A) in a gas or liquid. In the metric system pressure is measured in newtons per square meter (N/m^2); the SI unit is the *Pascal* [1 Pa = 1 N/m^2]. Standard atmospheric pressure at sea level (1 atm) is 101 kPa (1.01×10^5 Pa; 14.7 lb/in.^{2}; 760 mm Hg) and the pressure in a bicycle tire may be as high as 620 kPa (90 lb/in.^{2}). Though SI units are not in common use in medicine in the United States, they are used here as throughout the book. The common method of indicating pressure in medicine in the United States is by the height of a column of mercury (mm Hg). To convert from one system to the other, remember that 1 mm Hg = 0.133 kPa. For example, a peak (systolic) blood pressure reading of 120 mm Hg indicates that a column of mercury of this height has a pressure at its base equal to the patient's systolic blood pressure (see Chapter 8, *The Physics of the Cardiovascular System*). Table 5.1 lists some of the common units used to measure pressure and gives atmospheric pressure in each system.

Table 5.1. Conversion Factors for Common Pressure Units

	Atmospheres	Pa	cm H$_2$O	mm Hg	lb/in.^{2} (psi)
Atmosphere	1	1.01×10^5	1033	760	14.7
Pa	0.987×10^{-5}	1	0.0102	0.0075	0.145×10^3
cm H$_2$O	9.68×10^{-4}	98.1	1	0.735	0.014
mm Hg	0.00132	133	1.36	1	0.0193
lb/in.^{2} (psi)	0.068	6895	70.3	51.7	1

The pressure P (Pa) under a column of liquid of height h can be calculated from $P = \rho g h$, where ρ is the density of the liquid, and g is the acceleration due to gravity. Since the density of mercury is 13.6×10^3 kg/m^3, and that of water is 1.0×10^3 kg/m^3, a column of water has to be 13.6 times higher than a given column of mercury in order to produce the same pressure. It is sometimes convenient to indicate pressure differences in the body in terms of the height of a column of water.

Since we live in a sea of air with a pressure of about 1 atm, it is easier to measure pressure relative to atmospheric pressure than to measure true, or absolute, pressure. For example, if the pressure in a bicycle tire is 400 kPa (~60 lb/in.^{2}), the absolute pressure is 400 + 100 kPa = 500 kPa (~75 lb/in.^{2}). We say that 400 kPa is the *gauge pressure*. Unless we indicate otherwise, all the pressures used in this chapter are gauge pressures.

There are a number of places in the body where the pressures are lower than atmospheric, or negative. For example, when we breathe in (inspire or inhale), the pressure in the lungs must be lower than atmospheric pressure or the air would not flow in. The lung pressure during inspiration is typically a few centimeters of water [a few centipascals (cPa)] negative. When a person drinks through a straw, the pressure in his mouth must be negative by an amount equal to the height of his mouth above the level of the liquid he is drinking. Other examples of negative pressure are discussed in Chapter 7, *Physics of the Lungs and Breathing*.

Table 5.2 lists some typical pressures in the body. The heart acts as a pump, producing quite high pressure (~13 to 18 kPa, or ~100 to 140 mm Hg) to force the blood through the arteries. The returning venous blood is at quite low pressure and, in fact, needs help to get from the legs to the heart. The failure of this return system in the legs often results in varicose veins. Pressure in the circulatory system is discussed in detail in Chapter 8, *Physics of the Cardiovascular System*. In the current chapter we discuss some of the other pressure systems of the body and high pressure (hyperbaric) oxygen therapy.

Table 5.2. Typical Pressures in the Body

	Typical Pressure	
	kPa	mm Hg
Arterial Blood Pressure		
Maximum (systole)	13–18	100–140
Minimum (diastole)	8–12	60–90
Venous blood pressure	0.4–0.9	3–7
Great veins	<0.1	<1
Capillary blood pressure		
Arterial end	4	30
Venous end	1.3	10
Middle ear pressure	<0.1	<1
Eye pressure—aqueous humor	2.6	20
Cerebrospinal fluid pressure		
in brain (lying down)	0.6–1.6	5–12
Gastrointestinal	1.3–2.6	10–20
Intrathoracic pressure		
(between lung and chest wall)	−1.3	−10

5.1

PROBLEM

What height of water will produce the same pressure as 120 mm Hg? [Answer: 1.63 m H_2O]

5.2

PROBLEM

Calculate the pressure in Pa and in mm Hg equal to a pressure of 20 cm H_2O. [Answer: 1.96×10^3 Pa; 14.7 mm Hg]

5.3

PROBLEM

Assume you are a scuba diver preparing for a 10 m dive into salt water (density = 1.04×10^3 kg/m^3).

(a) What absolute pressure and what gauge pressure will you experience? [Answer: Absolute P: 2×10^4 Pa (2 atm), or 200 kPa; Gauge P: 1×10^5 Pa (1 atm) or 100 kPa]

(b) Assume your lungs have a maximum volume of 5×10^{-3} m^3 (5 liters). What will happen to that volume at a depth of 10 m? [Answer: Reduced to 2.5×10^{-3} m^3 (2.5 liters)]

(c) Suppose you cannot equalize the pressure in your middle ear. What will happen during the dive?
[Answer: At some depth, your eardrum will rupture.]

5.4

PROBLEM

If the density of air at sea level is 1.3 kg/m^3, calculate the pressure difference in Pa and in mm Hg between the bottom and top of a building 30 m tall (8 stories). [Answer: 3.8×10^2 Pa; 2.9 mm Hg]

5.1 Measurement of Pressure in the Body

The classical method of measuring pressure is to determine the height of a column of liquid that produces a pressure equal to the pressure being measured. An instrument that measures pressure by this method is called a *manometer*. A common type of manometer is a fluid-filled, U-shaped tube connected to the pressure to be measured (Fig. 5.1). The fluid levels in the vertical columns change until the difference in the levels is equal to the pressure difference. This type of manometer can measure both positive and negative gauge pressures. The fluid used is usually mercury, but water or other low density fluids can be used when the pressure to be measured is relatively small.

The most common clinical instrument used in measuring pressure is the *sphygmomanometer*, which measures blood pressure. Two types of pressure gauges are used in sphygmomanometers. In a mercury manometer, the pressure is indicated by the height of a column of mercury inside a glass tube. In an aneroid manometer, the pressure changes the shape of a sealed flexible container, which causes a needle to move on a dial.

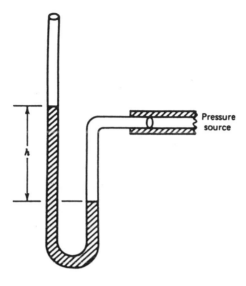

Figure 5.1. A U tube manometer for measuring pressure P. P can be expressed as the height of the fluid, h (e.g., in mm Hg or cm H_2O), or it can be expressed in conventional units of force per unit area by using $P = \rho g h$, where ρ is the density and g is the acceleration due to gravity.

Some parts of the body can act like crude pressure indicators. For example, a person going up or down in an elevator or an airplane is often aware of the change in atmospheric pressure on the ears. When one swallows, the Eustachian tubes can open momentarily and the pressure in the middle ear equalizes to the outside pressure; we may feel our eardrums "pop" as a result. Another qualitative pressure indicator is the size of the veins on the back of the hand. As a hand is raised slightly above the level of the heart, these veins become smaller due to the lower venous blood pressure (see Chapter 8).

5.5

PROBLEM

The venous pressure near the heart is typically 0.7 kPa (5 mm Hg). Describe a method to estimate this pressure.

5.6

PROBLEM

Positive pressure is used in blood transfusions. Suppose a container of blood is placed 1 m above a vein where the venous pressure is 0.3 kPa (2 mm Hg); if the density of blood is 1.04×10^3 kg/m^3, what is the net pressure acting to transfer the blood into the vein? [Answer: 9.9 kPa (74.5 mm Hg)]

5.7

PROBLEM

Negative pressure or suction is often used to drain body cavities. In the drainage arrangement for the gastrointestinal region, the negative pressure supplied to the collection bottle is –13 kPa (–100 mm Hg) and the top of the tube is 0.37 m above the end of the tube in the body. Make a sketch of this situation, and find the negative pressure at the lower end of the tube. [Answer: –69 mm Hg or –9.1 kPa]

5.2 Pressure Inside the Skull

The brain contains approximately 150 cm^3 of *cerebrospinal fluid* (CSF) in a series of interconnected openings called ventricles (Fig. 5.2). Cerebrospinal fluid is generated inside the brain and flows through the ventricles into the spinal column and eventually into the circulatory system. One of the ventricles, the aqueduct, is especially narrow. If at birth this opening is blocked for any reason, the CSF is trapped inside the skull and increases the internal pressure. The increased pressure causes the skull to enlarge. This serious condition, called *hydrocephalus* (literally, waterhead), is a moderately common problem in infants. However, if the condition is detected soon enough, it can often be corrected by surgically installing a by-pass drainage system for the CSF.

It is not convenient to measure the CSF pressure directly. One rather crude method of detecting hydrocephalus is to measure the circumference of the skull just above the ears. Normal values for newborn infants are from 0.32 to 0.37 m, and a larger value may indicate hydrocephalus. Another qualitative but more sensitive method of detection, *transillumination*, makes use of the light-scattering properties of the rather clear CSF inside the skull.

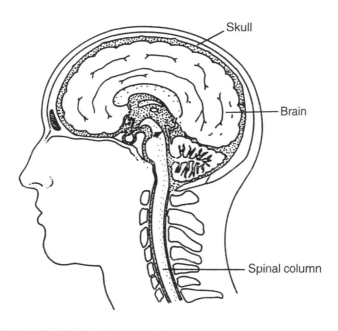

Skull

Brain

Spinal column

Figure 5.2. A cross section of the brain showing the location of the cerebrospinal fluid (shaded area) and the aqueduct (arrow). The fragile brain is supported and cushioned by this fluid.

5.3 Pressure in the Eye

The clear fluids in the eyeball (the aqueous and vitreous humors) that transmit the light to the retina (the light-sensitive part of the eye) are under pressure and maintain the eyeball in a fixed size and shape. The dimensions of the eye are critical to good vision—a change of only 10^{-4} m (0.1 mm) in its diameter has a significant effect on the clarity of vision. If you press on your eyelid with your finger you will notice the resiliency of the eye due to the internal pressure. The pressure in normal eyes ranges from 1.6 to 3 kPa (12 to 23 mm Hg).

The fluid in the front part of the eye, the aqueous humor, is mostly water. The eye continuously produces aqueous humor and a drain system allows the surplus to escape. If a partial blockage of this drain system occurs, the pressure increases and the increased pressure can restrict the blood supply to the retina and thus affect the vision. This condition, called *glaucoma*, produces tunnel vision in moderate cases and blindness in severe cases.

Early physicians estimated the pressure inside the eye by "feel" as they pressed on the eye with their fingertips. Now pressure in the eye is measured with several different instruments, called *tonometers*, that measure the amount of indentation produced by a known force. Tonometers are sometimes calibrated in arbitrary units rather than in kPa or millimeters of mercury. The tonometers currently being used are described in Chapter 12, *Physics of Eyes and Vision*.

5.4   Pressure in the Digestive System

The body has an opening through it. This opening, the digestive tract, is rather tortuous; it extends over 6 m from the mouth to the anus. Most of the time it is closed at the lower end and has several other restrictions. Figure 5.3 shows schematically the valves and sphincters (circular muscles) of the digestive tract, which open for the passage of food, drink, and their by-products. The valves are designed to permit unidirectional flow of food. With some effort it is possible to reverse the flow, such as during vomiting (emesis).

The pressure is greater than atmospheric in most of the gastrointestinal (GI) system. However, in the esophagus, the pressure is coupled to the pressure between the lungs and chest wall (*intrathoracic pressure*); this is usually less than atmospheric pressure. The intrathoracic pressure is sometimes determined by measuring the pressure in the esophagus.

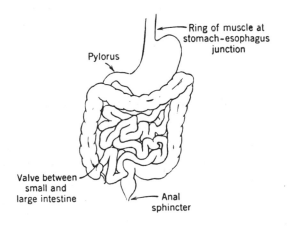

Figure 5.3. The valves and sphincters of the digestive tract.

During eating, the pressure in the stomach increases as the walls of the stomach are stretched. However, since the volume increases with the cube of the radius, R^3, while the tension (stretching force) is proportional to R, the increase in pressure is very slow. A more significant increase in pressure is due to air swallowed during eating. Air trapped in the stomach causes burping or belching. This trapped air in the stomach is often visible on an x-ray of the chest.

In the gut, gas (*flatus*) generated by bacterial action increases the pressure. (Flatus is produced in even the most cultured people!) External factors such as belts, girdles, flying, and swimming affect the gut pressure.

One valve, the *pylorus*, prevents the flow of food back into the stomach from the small intestine. Occasionally a blockage forms in the small or large intestine and pressure builds up between the blockage and the pylorus. If this pressure becomes great enough to restrict blood flow to critical organs, it can cause death. *Intubation*, the passing of a hollow tube through the nose, stomach, and pylorus, is usually used to relieve the pressure. If intubation does not work, it is necessary to relieve the pressure surgically. However, the high pressure greatly increases the risk of infection because the trapped gases expand rapidly when the incision is made. This risk can be reduced if surgery is performed in an operating room in which the external pressure is greater than the pressure in the gut—a hyperbaric operating room.

The pressure in the digestive system is coupled to that in the lungs through the flexible diaphragm that separates the two organ systems. When it is necessary or desirable to increase the pressure in the gut, such as during defecation, a person takes a deep breath, closes off the lungs at the *glottis* (vocal folds), and contracts the abdominal muscles. This increased pressure also reduces the flow of venous blood to the right side of the heart.

5.5 Pressure (Stress) in the Skeleton

The highest pressures in the body, which are more usually referred to as stresses, are found in the weight-bearing bone joints. When all the weight is on one leg, such as when walking, the stress in the knee joint may be more than 10^6 Pa (10 atm)! If it were not for the relatively large area of the joints, the stress would be even higher (Fig. 5.4). Since stress is defined as force per unit area, for a given force the stress is reduced as the area is increased.

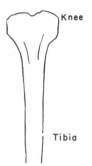

Figure 5.4. The surface area of a bone at the joint is greater than its area either above or below the joint. The larger area at the joint distributes the force, thus reducing the pressure.

Healthy bone joints are better lubricated than the best man-made bearings. If a conventional lubricant were used in a joint, it would be squeezed out and the joint would soon be dry. Fortunately, the joint is such that the higher the stress, the better the lubrication. Joint lubrication is discussed in more detail in Chapter 4, *Physics of the Skeleton*. Bone has adapted in another way to reduce stress. The finger bones are flat rather than cylindrical on the gripping side, and the force is spread over a larger surface; this reduces the stress in the tissues over the finger bones (Fig. 5.5).

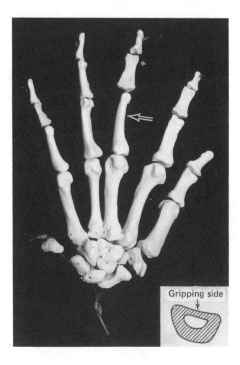

Figure 5.5. The finger bones rotated 90° to show the flat surface (arrow) used for gripping (see cross section in inset). This flat surface reduces the pressure on the tissues over the bones when we carry something heavy like a suitcase.

5.8

PROBLEM

Approximately 50% of your body weight lies above the fifth lumbar vertebra. If the cross-sectional area of that vertebra is taken to be about 3×10^{-3} m^2 (30 cm^2), what is the stress at that point in your spinal column? [Answer: 1.1×10^5 N/m^2]

5.6 Pressure in the Urinary Bladder

One of the most noticeable internal pressures is the pressure in the bladder resulting from an accumulation of urine. Figure 5.6 shows the typical pressure-volume curve for the bladder, which stretches as the volume increases. One might naively expect the rise in pressure to be proportional to the volume. However, as noted previously in the case of the stomach, for a given increase in radius R the volume increases as R^3 while the pressure only increases as R^2. This relationship largely accounts for the relatively low slope of the major portion of the pressure-volume curve in Fig. 5.6. For adults, the typical maximum volume in the bladder before voiding is 500 ml (0.5L). At some pressure, typically 3 kPa (or 30 cm H$_2$O), the *micturition* ("gotta go") reflex occurs. The resulting sizable muscular contraction in the bladder wall produces a momentary pressure of up to 15 kPa (150 cm H$_2$O). Normal voiding

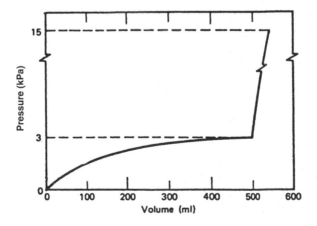

Figure 5.6. The typical pressure-volume relationship in the urinary bladder (cystometrogram).

pressure is fairly low, 2–4 kPa, (20–40 cm H_2O), but for men who suffer obstruction of the urinary passage as the result of an enlarged prostate the pressure required to affect voiding may be over 10kPa, 100 cm H_2O. This often leads to urine retention, which requires a catheter to empty the bladder.

The pressure in the bladder can be measured by passing a catheter with a pressure sensor into the bladder through the urinary passage (urethra). In direct *cystometry* the pressure is measured by means of a needle inserted through the wall of the abdomen directly into the bladder (Fig. 5.7). This technique gives information on the function of the exit valves (sphincters) that cannot be obtained with the catheter technique.

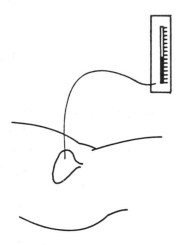

Figure 5.7. In direct cystometry a needle is passed through the wall of the abdomen directly into the bladder.

Bladder pressure increases during coughing, straining, and sitting up. During pregnancy, the weight of the fetus over the bladder increases the bladder pressure and causes frequent urination. A stressful situation may also produce a pressure increase; studying for exams often results in many trips to the toilet due to "nerves."

5.7 Pressure Effects While Diving

Pressure changes while diving do not greatly affect most of the body since it is composed primarily of solids and liquids which are nearly incompressible. However, there are gas cavities in the body where pressure changes can produce profound effects. To understand why, we must recall Boyle's law: for a fixed quantity of gas at a constant temperature, the product of the absolute pressure and volume is constant (PV = constant). That is, if the absolute pressure is doubled, the volume is halved.

The middle ear is one air cavity that exists within the body (see Fig. 10.11). For comfort the pressure in the middle ear should nearly equal the atmospheric pressure outside the eardrum. Equalization is produced by air flowing through the *Eustachian tube*, which is usually closed except during swallowing, chewing, and yawning. When diving, many people have difficulty obtaining pressure equalization and feel pressure in their ears. A pressure differential of 17 kPa (120 mm Hg) across the eardrum, which can occur in about 1.7 m (5.3 ft) of water, can cause the eardrum to rupture. Rupture can be serious since cold water in the middle ear can affect the vestibular or balance mechanism and cause nausea and dizziness. One method of equalization used by a diver is to raise the pressure in the mouth by holding the nose and trying to blow out; as the pressure equalizes the diver can often "hear" both ears "pop." (This same method can be attempted to relieve discomfort from pressure changes which occur in travel by airplane when you are descending.)

A less serious condition is sinus squeeze. During a dive the pressure in the sinus cavities in the skull usually equalizes with the surrounding pressure. If a diver has a cold, the sinus cavities may become closed off and not equalize, causing pain. Another pressure effect is pain during and after dives from small volumes of air trapped beneath fillings in the teeth. Eye squeeze can occur if goggles are used instead of a facemask; with a facemask the air exhaled from the lungs increases the pressure over the eyes as the descent is made.

If a scuba diver at a depth of 10 m holds her/his breath and comes to the surface, the air volume will expand by a factor of two and thus cause a serious pressure rise in the lungs. If the lungs are filled to capacity, an ascent of only 1.2 m (4 ft) while holding your breath can cause serious lung damage. All scuba divers learn during training to avoid breath-holding during ascent and to exhale continuously if a rapid ascent is necessary.

The pressure in the lungs at any depth is greater than the pressure in the lungs at sea level. This means that the air in the lungs is more dense underwater and that the partial pressures of all the air components are proportionately higher. The higher partial pressure of oxygen causes more oxygen molecules to be transferred into the blood, and oxygen poisoning results if the partial pressure of oxygen gets too high. (Fig. 5.8). Usually, oxygen poisoning occurs when the partial pressure of oxygen is above 80 kPa (when the absolute air pressure is above 4 atm), or at a depth greater than 30 m (100 ft).

Breathing air at a depth below 30 m is also dangerous because it may result in excess nitrogen in the blood and tissues. This can produce two

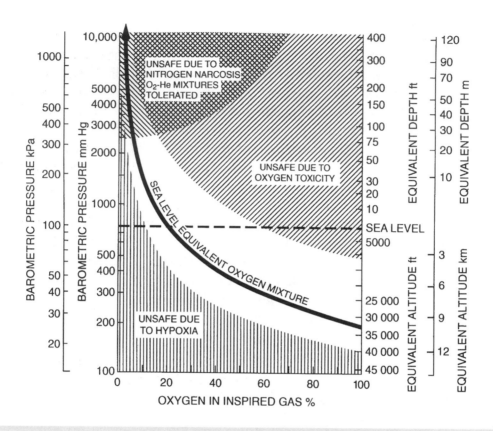

Figure 5.8. At the normal 20% O_2 mixture and at sea water depth greater than 30 m (100 ft) both oxygen poisoning and nitrogen narcosis can occur. Note that the higher pressures are at the top of the graph and that the pressure decreases as the altitude increases. At an altitude of about 4600 m (~15,000 ft) hypoxia (a shortage of oxygen in the tissues) can occur. (Modified from C. E. Billings in J. F. Parker, Jr., and V. R. West (Eds.), *Bioastronautics Data Book*, 2nd. Ed., National Aeronautics and Space Administration, Washington, DC, 1973, p. 2).

serious problems: nitrogen narcosis, which is an intoxication effect (Fig. 5.8), and the bends, or decompression sickness, which is an ascent problem. While oxygen is transported primarily by chemical attachment to the red blood cells, nitrogen is dissolved in the blood and tissues. According to Henry's law, the amount of gas that will dissolve in a liquid is proportional to the partial pressure of the gas in contact with the liquid. As a diver goes deeper, the pressure of the air and thus the partial pressure of nitrogen increase. As a result, more nitrogen is dissolved in the blood and from there into the tissues. The same is true of O_2, but it is used by the tissues. When the diver ascends, the extra nitrogen in the tissues must be removed via the blood and the lungs. The removal is a slow process and if the diver ascends too fast, bubbles form in the tissues and joints. The bends are quite painful. Stricken divers are usually re-compressed in a chamber; the pressure in the chamber is slowly decreased so that the nitrogen can be removed from the tissues via the blood and the lungs.

Other problems can occur during ascent. One of the membranes that separate air and blood in the lung can burst, allowing air to go directly into the bloodstream (air embolism). Air can also become trapped under the skin around the base of the neck or in the middle of the chest. In addition, *pneumothorax* (lung collapse) can result if air gets between the lungs and the chest wall (see Chapter 7, *Physics of the Lungs and Breathing*). These problems are best treated by a physician.

5.9

PROBLEM

What volume of air at an atmospheric pressure of 100 kPa (1.0×10^5 N/m^2) is needed to fill a 1.42×10^{-2} m^3 (14.2 liter, 0.5 ft^3) scuba tank to a pressure of 1.4×10^4 kPa (140 atm, 2100 lb/in.^{2})? [Answer: 2×10^3 liters = 72 ft^3]

5.10

PROBLEM

Since at sea level a diver uses about 14.2 liters (0.5 ft^3) of air per minute during moderate activity, the tank in Problem 5.9 would last about 144 min. How long would the tank last at a depth of 10 m (33 ft) where the pressure is increased by 1 atmosphere, assuming the same volume use rate? [Answer: 72 min]

5.11

PROBLEM

Suppose you are a deep-sea diver in a dive to 30 m.
(a) What absolute and gauge pressures will you experience?
    [Answer: 300 kPa gauge (3 atm); 400 kPa absolute (4 atm)]
(b) What will be your rate of air consumption compared to that at sea level? [Answer: 4 times as fast]

5.8 Hyperbaric Oxygen Therapy (HOT)

The body normally lives in an atmosphere that is about one-fifth oxygen and four-fifths nitrogen. In some medical situations it is beneficial to increase the proportion of oxygen in order to provide more oxygen to the tissues. Oxygen tents are often used for this purpose. To greatly increase the amount of oxygen, medical engineers have constructed special high pressure (*hyperbaric*) oxygen chambers. Some are just large enough for a patient, while others are large enough to serve as operating rooms. Gas gangrene is a disease that killed more than half of its victims before *hyperbaric oxygen therapy* (HOT) was developed. Since the bacillus that causes gas gangrene cannot survive in the presence of oxygen, almost all gas gangrene patients treated with HOT are cured without the need for amputation—previously the best method of treatment.

In carbon monoxide poisoning the red blood cells cannot carry oxygen to the tissues because the carbon monoxide keeps the hemoglobin from transporting oxygen. The presence of even a few carbon monoxide molecules on a red blood cell greatly reduces the ability of the cell to transport oxygen. Normally the amount of oxygen dissolved in the blood is about 2% of that carried on the red blood cells. With HOT, the partial pressure of oxygen can be increased by a factor of 15, permitting enough oxygen to be dissolved to fill the body's needs. Many victims of carbon monoxide poisoning are saved with this technique.

Like many new developments in medicine, hyperbaric oxygen therapy brought with it new problems. The oxygen atmosphere makes fire a much greater hazard—three astronauts died in the pure oxygen atmosphere on a U.S. spaceship during preliminary tests in 1967. Another problem is the risk of rupture of the tank due to the high pressures used. Such a rupture occurred on at least one occasion, seriously injuring the patient and the physician in attendance. However, physical dangers such as these are usually easier to evaluate and avoid than biological dangers (e.g., air pollution), which are often poorly understood.

6

Osmosis and the
Kidneys

Russell K. Hobbie[1]

Two physical processes are important in the transport of nutrients and wastes in the body: *diffusion* and *solvent drag*. For example, carbon dioxide produced in cellular metabolism diffuses from the cells to the capillaries. Once dissolved in the blood, it is carried along in the flowing blood by solvent drag. This chapter describes how substances are transported by these two processes, with examples of how the processes work in the blood and lymphatic systems and in the kidney.

6.1 How Substances Are Transported in Fluids

6.1.1 Solvent Drag

Imagine a long tube filled with flowing water. The flow rate is i m^3 s^{-1}. If a solute such as glucose is dissolved in the water at concentration C

---

[1]This chapter is based on R. K. Hobbie, *Intermediate Physics for Medicine and Biology*, 3rd Edition, Springer-Verlag New York, Inc., 1997. Reproduced with permission of Springer-Verlag New York, Inc.

127

particles m^{-3}, it will be carried along with the water. The flow of solute particles past a point in the tube is Ci particles s^{-1}. This process is called solvent drag; the solute particles are dragged along by the solvent. This is the process by which red and white blood cells and dissolved chemicals like glucose and carbon dioxide are carried through blood vessels.

6.1.2 Diffusion

Suppose now that the water in the tube is not flowing. If the glucose concentration is uniform along the tube, it will not flow either. If, however, there is glucose at one end of the tube but not the other, it will gradually fill the tube to a uniform concentration, even though the water is not flowing.

The reason is that the water and glucose molecules are not at rest, even though the water is still, which means that it is not moving *on average*. Each molecule is moving helter-skelter, back-and-forth, with an average speed which increases with the temperature. Only at absolute zero would it be perfectly still. A molecule does not move very far (less than a molecular diameter) before it collides with another molecule and changes direction. Its motion is completely random; it is equally likely to travel to the right as to the left. We often say that the molecules are undergoing a *random walk*. When the water is flowing, a drift of all the moleces in one direction is superimposed on this random motion in all directions.

Consider an imaginary boundary at some point along the tube. In a short time interval about half of the glucose molecules immediately to the left of this point will travel briefly to the right and cross the boundary. The other half will be traveling to the left, since random motion to the left or to the right is equally probable. At the same time, about half of the glucose molecules immediately to the right of this point will travel left and cross the boundary. If the concentration of glucose molecules on the left is the same as on the right, the net result is no change of glucose concentration; as many molecules moved from left to right as moved from right to left.

Suppose, however, that two adjoining regions have different glucose concentrations, as shown in Figure 6.1. Any molecule is equally likely to move to the right or left. Half of those on the left (region A) will move right, and half of those on the right (region B) will move left. There will be a net movement of glucose from left to right, not because glucose molecules prefer to travel to the right, but because there were more on the left to wander to the right than there were on the right to

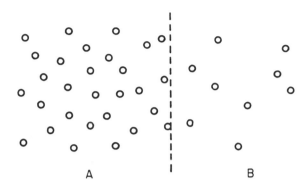

Figure 6.1. An example of diffusion. Each molecule in region A or B can wander with equal probability to the left or to the right. There are more molecules at A to wander to the right than there are at B to wander to the left. The result is a net flow of molecules from A to B. From R. K. Hobbie, *Intermediate Physics for Medicine and Biology*, 3rd Ed., Springer-Verlag New York, 1997. Reproduced with permission of Springer-Verlag New York.

wander to the left. This process is called *diffusion*. Diffusion requires a change in concentration with distance, or the existence of a *concentration gradient*.

6.1.3 Bulk Flow vs. Diffusion

An important distinction between bulk flow and diffusion is sometimes overlooked. The distinction is shown in Figure 6.2. The open circles represent water molecules, and the solid circles are solute particles. The water molecules are in contact with one another. If one water molecule moves, it no longer touches its neighbors, and they sense its motion. Thus, the water molecules move collectively. This is bulk flow. In a tube such as a blood vessel, the bulk flow is caused by a pressure difference between the ends of the tube.

Each solute molecule, on the other hand, is surrounded by water molecules, not by other solute molecules. As a result, it does not detect the motion of other solute molecules. It is influenced only by the motion of the neighboring water molecules. Motion of the surrounding water molecules during bulk flow causes the solute molecule to be dragged along, which was defined as *solvent drag*. If there is no bulk flow, each solute molecule undergoes its own random motion, independent of the other solute molecules. If there is a concentration

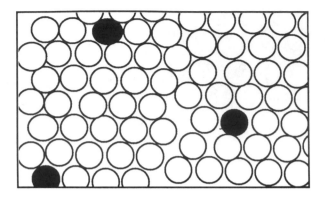

Figure 6.2. A collection of water and solute molecules (open and closed circles). The water molecules are surrounded by other water molecules and move together. The solute molecules are isolated and move independently of other solute molecules, though they experience solvent drag if the water molecules move.

gradient of solute molecules, there will be diffusion. There can be solvent drag and diffusion at the same time.

6.2 How Substances are Transported Through Membranes

6.2.1 Diffusion and Solvent Drag

Most volumes of biological interest are bounded by membranes. Every cell is surrounded by a membrane. Nearly all substructures within a cell are bounded by membranes. The wall of a capillary is a membrane. Membranes often allow some substances to pass through but not others. The membrane is *permeable* to substances which can pass through and *impermeable* to those that cannot. A membrane that allows some substances to pass through but not others is *semipermeable*.

Many biological membranes contain pores. The simplest pore would pass straight through the membrane as shown in Fig. 6.3a. The pore usually follows a longer path, as in Fig. 6.3b. In the capillary wall the pores consist of regions between endothelial cells on the inside of the membrane. There is excellent evidence that cell membranes contain pores.

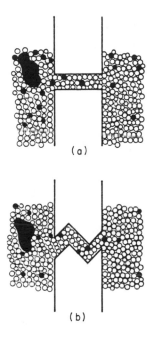

Figure 6.3. A membrane pierced by pores. (a) A straight pore. (b) A pore following a tortuous path. R. K. Hobbie, *Intermediate Physics for Medicine and Biology*, 3rd Ed., Springer-Verlag New York, 1997. Reproduced with permission of Springer-Verlag New York.

Pores provide one simple way to make a semipermeable membrane. A membrane pierced by pores will be permeable to molecules which are small enough to pass through the pore and impermeable to those that are too large. Water flows through the pores (bulk flow) if there is a pressure difference across the membrane. Solute molecules which are small enough to enter the pore will be carried by solvent drag if there is bulk flow. They will also diffuse if there is a concentration gradient along the pore. In most cases the flow of a solute is a combination of both effects.

6.2.2 Osmotic Pressure

Consider two compartments of fluid (say, water) separated by a semipermeable membrane. If the fluid can pass through the membrane, there is no flow when the pressure on both sides is the same. If we now introduce on the left some solute molecules which cannot pass through the membrane, a very interesting effect occurs. As this is done, the volume on the left is increased, or some water molecules are removed, to keep the pressure the same. We discover that, if the pressure is the same on both sides, water flows from right to left. We can make the flow of water zero if we increase the pressure on the left. The amount by which it must be increased is the *osmotic pressure* of the solute.

There are two ways to visualize the need for an increased pressure to prevent flow. First, consider just the water molecules. The introduction of the impermeant solute molecules reduces the number of water molecules per unit volume on the left. Even though the water molecules are moving collectively, and the phenomenon is not diffusion, there are more molecules per unit volume on the right than on the left. To prevent motion from right to left, the pressure on the left must be increased.

Second, we can consider a case in which the pressure has been increased on the left so that there is no flow through the pores in the membrane. The pressure on the membrane is caused by repeated collisions of the fluid molecules (both water and solute) with the membrane. The impermeant solute molecules, some of which are shown in Figure 6.3, are larger than the pore. They collide with the membrane, and, along with the smaller molecules, contribute to the total pressure. However, they cannot enter the pore. The membrane prevents them from striking the water molecules just inside the mouth of the pore. As a result, the pressure inside the pore is less than on the left. This permits water to flow into the side with the solute.

The *driving pressure* p_d on each side of the membrane is equal to the total pressure (measured, for example, with a gauge) minus the osmotic pressure, π: $p_d = p_{total} - \pi$. The difference in driving pressure across the membrane causes bulk flow through the pores. Some books talk, incorrectly, about the diffusion of water. Since the water molecules are in contact with one another, they do not undergo random walk, moving independently of one another (as described in section 6.1.2), which is the hallmark of diffusion. They flow by bulk flow in exactly the same way, whether the difference in driving pressure is caused by a total pressure difference or an osmotic pressure difference.

Under certain conditions when the viscosity of the fluid is important, the total flow is proportional to the fourth power of the tube radius. That is, for a given pressure drop in a tube of given length, increasing the radius by a factor of two increases the flow rate $2^4 = 16$ times. The flow rate for a given difference in driving pressure is exactly the same, whether the difference is caused by a difference in osmotic pressure or in total pressure, or some combination.

6.2.3 Active Transport—Going the Wrong Way

Sometimes substances can move across a membrane from a region of lower to higher concentration. Such movement cannot be due to random motion. Rather, it is an instance of active transport, in which chemical free energy is expended to pump the substance up the concentration gradient. (The situation is analogous to a hill. It is possible to fall downhill, but energy is required to move uphill.)

6.3 Regulation of the Interstitial Fluid

As blood flows through the capillaries, oxygen and nutrients leave the blood and go to the cells (see Chapter 8, *Physics of the Cardiovascular System*). Waste products leave the cells and enter the blood. Diffusion is the main process that accomplishes this transfer. The length of a typical capillary is 1 mm. The capillaries are about the diameter of a red cell, 7 μm (7×10^{-6} m); the red cells therefore squeeze through the capillaries single file. They are carried in the plasma, which consists of water, electrolytes, small molecules such as glucose and carbon dioxide, and large protein molecules. All of the plasma except the large protein molecules can pass through the capillary wall. There are small slits between the cells making up the capillary wall that function as pores.

Outside the capillaries is the *interstitial fluid*, which bathes the cells. The concentration of protein molecules in the interstitial fluid is much less than in the capillaries. Osmotic pressure is an important factor in determining the pressure in the interstitial fluid and therefore the flow of nutrients and water is through the capillary wall. The following values are typical for the osmotic pressure inside and outside the capillary.

Inside capillary: $\pi_i = 3700$ Pa (28 mm Hg)

Outside capillary, interstitial fluid: $\pi_o = 600$ Pa (4.5 mm Hg)

Measurements of the total pressure in the interstitial fluid are difficult, but the value seems to be about −800 Pa (−6 mm Hg). The tissue has to be somewhat rigid or the pressure in the interstitial fluid would be atmospheric. It is like a slightly evacuated can, not a balloon. It is maintained below atmospheric pressure (taken here to be zero) by the rigidity of the tissues. The driving pressure of water and small molecules outside, p_{do}, is therefore

$$p_{do} = p_o - \pi_o = -800 - 600 = -1400 \text{ Pa } (-10.5 \text{ mm Hg}) \quad (6.1)$$

The total pressure within the capillary drops from the arterial to the venous end, causing blood to flow along the capillary. A typical value at the arterial end is 3300 Pa (25 mm Hg); at the venous end it is 1300 Pa (10 mm Hg). If the pressure drop along the capillary is linear, the total pressure vs. position is as plotted in Figure 6.4a. Subtracting from this the osmotic pressure due to the large molecules gives the curve for the driving pressure p_{di}, which is also plotted in Figure 6.4a. Figure 6.4b shows the total and driving pressures in the interstitial fluid. Figure 6.4c compares the driving pressure inside and outside. The pressure is larger inside in the first half of the capillary and is larger outside in the second half. The result is an outward flow of plasma (without the large proteins) through the capillary wall in the first half and an inward flow in the second half. There is a very slight excess outward flow. This fluid returns to the circulation through the lymphatic system, a system of vessels and lymph nodes which parallels the veins and enters the venous circulation near the heart.

There are three ways that the balance of Figure 6.4c can be disturbed. Each can give rise to collection of fluid in the tissue and localized swelling called *edema*. The first is a higher pressure along the capillary. This can happen in heart failure. The left heart pumps blood from the lungs through the body. If it fails to pump well, blood will accumulate in the blood vessels of the lungs, leading to a pressure buildup in the lungs which causes pulmonary edema. Failure of the right heart, which pumps blood from the peripheral circulation through the lungs, can lead to edema in the feet and legs.

The second cause of edema is a reduction in osmotic pressure of the plasma because of a low protein concentration (*hypoproteinemia*). This can happen in malnutrition, in kidney disease in which protein is lost in the urine, or in liver disease. In each case it leads to severe edema. Diseases of the liver can block the return of venous blood from the intestine to the heart, leading to abdominal edema called *ascites*.

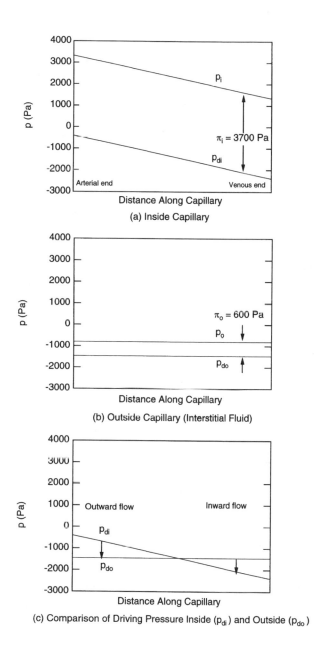

Figure 6.4. Total and driving pressures inside and outside a capillary. (a) Inside. (b) Outside. (c) Comparison of driving pressure inside and outside.

The third is an increased permeability of the capillary wall to large molecules, which effectively reduces the osmotic pressure. The swelling from a bad bruise is an example of this kind of swelling.

6.4 The Kidney

The kidneys excrete much of the body's metabolic waste products—except carbon dioxide and some water which leave through the lungs. They also regulate the concentration of most chemicals in the blood plasma. Each kidney contains over 1 million *nephrons*. Each nephron is a complete urine-forming unit. Figure 6.5 shows the kidneys and the ureters through which urine flows to the urinary bladder. Figure 6.6 shows a magnified view of a nephron.

Figure 6.7 shows the essential functioning parts of the nephron. Blood from a renal artery passes first by a membrane in the *glomerulus*, where a large amount of fluid—about 250 ml per minute (~1 cup)—passes through the basement membrane of the glomerulus. This process is called *filtration*. Careful measurements on dog kidneys using radioactively tagged solute molecules of different radii suggest that the filtration is by pores of 5 nm radius in the basement membrane. The filtration rate is controlled by valves which control the rate of blood flow through the

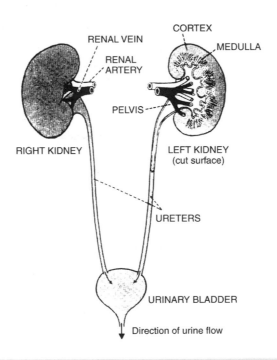

Figure 6.5. The kidneys, ureters, and urinary bladder. From Arthur C. Guyton, *Textbook of Medical Physiology*. Philadelphia, W. B. Saunders, 1976. Reproduced with permission of W. B. Saunders Company.

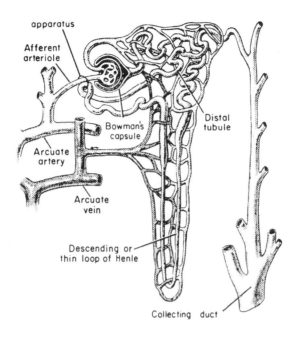

Figure 6.6. A single nephron. From Smith, *The Kidney: Structure and Functions in Health and Disease*, Oxford University Press, 1951. Used by permission of Oxford University Press.

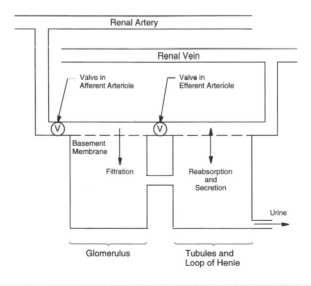

Figure 6.7. A schematic view of the nephron. The essential parts are the glomerulus, where initial filtration takes place, and the tubules, where reabsorption and secretion take place. The valves in the afferent and efferent arterioles regulate the pressure difference across the basement membrane of the glomerulus.

glomerulus and the pressure drop across the glomerular basement membrane. Substances with a molecular weight of 5000 or less pass easily through the membrane with the water. Most proteins, which have a molecular weight of 69,000 or more, do not pass through the pores and remain in the blood. The filtrate then passes through the tubules, where 99% of it is reabsorbed (if it were not reabsorbed, we would void 360 liters of urine per day). The other 1% passes into the collecting system as urine. Unwanted substances are not reabsorbed, so their concentration in the urine increases. Creatinine, a metabolic waste product, and sucrose are not reabsorbed at all. About half of the urea, a nitrogenous product of protein metabolism, is reabsorbed.

Osmotic pressure is important in both the initial filtration in the glomerulus and in the reabsorption in the tubules. Since protein molecules do not pass through the basement membrane, their osmotic pressure in the plasma is important. The filtration rate is proportional to the difference in driving pressure across the glomerular membrane, that is, the difference in $p - \pi$ on the two sides. For reabsorption in the tubules, many more molecules contribute to the osmotic pressure, since they are not reabsorbed. If the concentration of one of these smaller molecules in the tubules is unusually high, less water than normal will be reabsorbed and urine flow will be increased. This effect is called *osmotic diuresis*. Glucose is normally completely reabsorbed. However, if the level of glucose in the plasma becomes too high, such as in diabetes, the kidneys cannot reabsorb all of it. The glucose that remains in the tubules contributes to the osmotic pressure and therefore causes less water reabsorption than normal. Diabetics with a very high plasma glucose (>500 mg/dl) therefore produce abnormally high volumes of urine. Frequent urination and a great thirst to replace water are symptoms of uncontrolled diabetes.

Other substances such as sodium are reabsorbed by active transport. Certain cells lining the tubule use chemical reactions to reduce the sodium concentration inside the cells. This allows sodium from the tubule to diffuse in. Once inside, it is removed by active transport, which involves the expenditure of chemical energy to move the sodium ions to a region of higher concentration. Other substances for which active transport is important are glucose, calcium, potassium, and amino acids. Some substances, such as hydrogen and potassium ions, are actively transported into the tubule. This process is called *secretion*.

The *plasma clearance* or simply *clearance* expresses the ability of the kidney to remove various substances from the plasma. The rate of removal of the substance from the body (R molecules per second) is proportional to the concentration of the substance in the plasma (C molecules per m^3). This makes sense if we think about the kidney. Whatever processes take place there can only depend on the concentration of the substance in the blood, not on the total blood volume. If we were to double both the blood volume and the amount of some substance, thereby keeping the concentration the same, the transport processes in the kidney would continue to function at the same rate. It would then take twice as long to remove all the substance because there would be twice as much of it. The proportionality constant is the clearance, K, whose units are m^3/s^{-1}:

$$R \text{ (molecules/s)} = K \text{ (}m^3/s\text{) } C \text{ (molecules/}m^3\text{)} \qquad (6.2)$$

Equation 6.2 can also be written in terms of the mass of the removed particles, and in terms of concentrations per liter, and per hour instead of per second. The units of R are the units of K times the units of C; for example,

$$R \text{ (mg/hour)} = K \text{ (liter/hour) } C \text{ (mg/liter)} \qquad (6.3)$$

The value of K usually depends on the pressure differences, the osmotic pressure, and the chemistry of any active transport processes. An equivalent definition is "clearance is the volume of plasma from which the substance is completely removed per second."

6.5 The Artificial Kidney

It is possible to use dialysis to remove small solute molecules such as urea from the blood of patients with acute or chronic kidney disease. A portion of the patient's blood is circulated through tubes in a machine. The walls of these tubes are a semipermeable membrane immersed in the dialysis fluid. The membrane allows water, electrolytes, and urea molecules to diffuse through pores into the dialysis fluid. The electrolytes in the dialysis fluid are kept at the concentration desired in the patient's blood. The process takes several hours.

It is possible to use a membrane with larger surface area to speed up the process, but severe headaches result. The reason is that a membrane separates the blood in the central nervous system from the cerebrospinal fluid (CSF). Certain medium-sized molecules move slowly across this membrane. If these molecules are removed from the blood too rapidly during dialysis, their concentration is significantly higher in the cerebrospinal fluid than in the blood. The resulting osmotic pressure causes flow of water into the CSF, increasing pressure in the skull. The result is a headache.

6.1 PROBLEM

Blood flows through a vessel at a rate of 3 ml/s (1 milliliter = 10^{-6} m^3). It contains albumin, which has a molecular weight of 75,000, at a concentration of 4.5 gm per 100 ml.
(a) How many grams of albumin per second flow through the vessel?
    [Answer: 0.135 gm/s]
(b) What is the concentration of albumin, in molecules/m^3?
    [Answer: 3.6×10^{23} molecules/m^3]

6.2 PROBLEM

In certain cases the flow of fluid through a vessel is proportional to the fourth power of the diameter of the vessel (with the same pressure drop per unit length along the tube). If, in such a case, the diameter of the tube is increased by 25%, what is the increase in flow?
[Answer: 2.4 times]

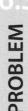

6.3 PROBLEM

The amount of water (in the blood and tissues) in a typical person is about 18 liters. Suppose that the concentration of some substance is 3 mg/liter. The clearance of this substance through the kidneys is 2 liter/hr. At what rate is the substance being removed from the body by the kidneys? [Answer: Equation 6.2 can also be written in terms of the mass of the removed particles, and in terms of concentrations per liter, and per hour instead of per second:
R (mg/hour) = K (liter/hour) C (mg/liter) = 2 (liter/hour) 3 (mg/liter) = 6 mg/hour.]

PROBLEM 6.4

Suppose that the concentration of the substance in Problem 6.3 has been reduced to 1 mg/liter. At what rate is the substance then being removed? Will all of the substance ever be removed? [Answer: Now the rate is 2 mg/hour. It will never all be removed because as the amount becomes less, the concentration becomes less, and the rate of removal becomes less.]

PROBLEM 6.5

The clearance of a drug from the body is 100 ml/hr. It is stored in 18 liters of body water. The desired concentration in the blood is 0.1 mg/liter. How much of the drug must be given each day to maintain this concentration? [Answer: Recall that rate of removal = R; clearance = K; concentration = C. In this case, we have $R = K\,C = 0.24$ mg day^{-1}. To keep the same concentration, this amount must be given per day.]

PROBLEM 6.6

Suppose that a semipermeable membrane separates two volumes. The total pressure in both volumes is the same. Solutes are arranged as shown:

Quantity	On the Left	On the Right
Solute 1, which can pass through the membrane, has the same concentration on both sides.	C_1	C_1
Solute 2, which cannot pass through the membrane, exists only on the right.	None	C_2

Describe whether solute 1 moves through the membrane, and if so, the mechanism by which it moves. [Answer: Because the total pressure is the same on both sides and there is an osmotic pressure on the right that does not exist on the left, there is bulk flow from left to right. Solute 1 will not move by diffusion, because the concentrations are the same on both sides. It will move from left to right by solvent drag. This depends on whether there is a significant volume flow from left to right.]

7

Physics of the Lungs and Breathing

The body is in many ways a very remarkable machine. It must have a source of energy, a method of converting the energy into electrical and mechanical forms, and a way of disposing of its by-products. In an automobile, the source of energy is gasoline; it is combined with air and burned in the cylinders to produce kinetic energy to drive the wheels, and its by-products of noxious gases and heat are disposed of through the exhaust and the radiator. In the body, the source of energy is food; it is processed in the digestive system and then combined with O_2 in the cells of the body to release energy. Its by-products are disposed of by four routes: (1) the non-digestible solid components are eliminated as feces, (2) water and some other by-products are carried away in the urine, (3) almost 0.5 kg of CO_2 is disposed of via the lungs each day, and (4) heat is dissipated from the body's surface.

The human "machine" really consists of trillions of very small "engines"—the living cells of the body. Each of these miniature engines must be provided with fuel, O_2, and a method of getting rid of the by-products. The blood and its vessels (cardiovascular system) serve as the transport system (Chapter 8). The lungs (pulmonary system) serve as the supplier of O_2 and the disposer of the main by-product, CO_2. The blood takes the O_2 to the tissues and removes the CO_2 from the tissues; it must come in close contact with the air in the lungs in order to exchange its load of CO_2 for a fresh load of O_2 (Fig. 7.1). The details of this process are discussed in Section 7.2.

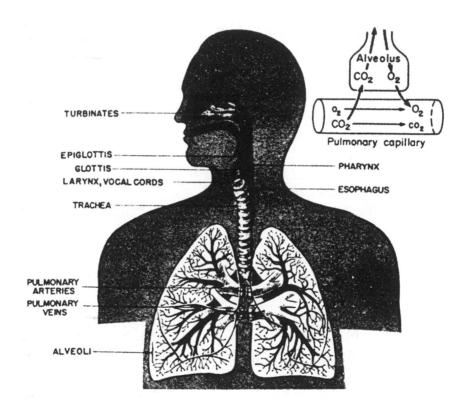

Figure 7.1. The major channels for air and blood in the respiratory system. The diagram illustrates schematically the exchange of O_2 and CO_2 between the air in an alveolus and the blood. (From Guyton, A. C., *Function of the Human Body*, 3rd Ed., © W. B. Saunders Company, Philadelphia, 1969, p. 222. Reprinted by permission.)

Because of the close cooperation and interactions between the cardiovascular and pulmonary systems, the actions of one system often affect the other. For example, during breathing the pressure on the major veins in the chest affects the return of blood to the heart. Often a disease of the lungs will produce heart symptoms and vice versa.

The lungs perform other physiologic functions in addition to exchanging O_2 and CO_2. One primary function is keeping the pH (acidity) of the blood constant. The lungs play a secondary role in the heat exchange of the body (Chapter 2) by warming and moisturizing the air we breathe in. Our breathing mechanisms provide a controlled flow of air for talking, coughing, sneezing, sighing, sobbing, laughing, sniffing, and yawning.

An important function of the breathing apparatus is voice production. Breathing patterns are markedly different during conversation. Since the voice is produced by a controlled outflow of air from the lungs, a person inhales rapidly and more deeply before speaking in order to have more time to produce voice sounds. The inhalation time is typically less than 20% of the breath cycle, and the amount inhaled is usually more than twice the usual volume when not talking.

A singer (especially an opera singer) inhales even more air in a short period of time to further minimize the inhalation part of the cycle. The airway resistance produced by the vocal cords causes a sizable pressure increase in the trachea. Thus the work involved in speaking and singing is considerably greater than the work of normal breathing. However, relatively little of the increased work goes into sound energy. The voice typically has a power of less than 1 mW. Voice production is discussed further in Chapter 10, *Sound and Speech*.

We breathe about 6 liters (6×10^{-3} m^3) of air per minute. Coincidentally, this is also about the volume of blood the heart pumps each minute. Men breathe about 12 times per minute at rest, women breathe about 20 times per minute, and infants breathe about 60 times per minute. We discuss in Section 7.6 the physical factors that affect the breathing rate.

The air we inspire is about 80% N_2 and 20% O_2. Expired air is about 80% N_2, 16% O_2, and 4% CO_2. (This relatively high percentage of O_2 in exhaled air is the reason that mouth-to-mouth resuscitation is practical and that blowing on a campfire helps get it started.) We breathe about 10 kg (22 lb) of air each day. Of this, the lungs absorb about 0.5 kg of O_2; about 400 liters of O_2), and release a slightly smaller amount of CO_2. We also saturate the air we breathe with water. When we breathe in dry air, our expired air carries away about 0.5 kg of water each day. (This moisture can be used to clean glasses.) In cold weather some of this moisture condenses and we see our breath. The air we breathe contains dust, smoke, air-borne bacteria, noxious gases, and so forth, which come into close contact with the blood. The large, convoluted surfaces of the lungs have a surface area of about 80 m^2 about one-half the area of a singles tennis court! The lungs have a greater exposure to the environment than any other part of the body including the skin. It is perhaps surprising that we don't have more diseases of the lungs. The importance of clean air is obvious.

Each time we breathe, a volume of about 0.5 liters containing ~10^{22} molecules of air enters our lungs. The total number of molecules in the earth's atmosphere is about 10^{44}. We thus take in $1/10^{22} = 10^{-22}$ of all the earth's air each time we breathe; in other words, for each molecule we breathe there are 10^{22} more in the earth's atmosphere. The earth's atmosphere is in constant motion, and over a period of centuries there has been thorough mixing of the gases. As a result, each breath, or 0.5 liter of air (10^{22} molecules), contains on the average one molecule that was present in any 0.5 liter of air centuries ago. An interesting way to think of this is that on the average each of our breaths contains one air molecule that was in a single breath of Archimedes, Aristotle, or any other famous, infamous, or unknown person who lived many years ago. Jesus Christ took approximately 150 million breaths in his lifetime; thus, one could expect that each of our breaths could contain about 150 million molecules breathed by Christ.

7.1 PROBLEM

Calculate the number of O_2 molecules absorbed by the body from a typical breath. [Hint: the percentage of O_2 per unit volume in inhaled air is 20% and in exhaled air is 16%.]
[Answer: 5×10^{20} O_2 molecules/breath]

7.1 The Airways

The principal air passages into the lungs are shown in Fig. 7.2. Air normally enters the body through the nose where it is warmed (if necessary), filtered, and moisturized. The moist surfaces and the hairs in the nose trap dust particles, bugs, and so forth. During heavy exercise, such as jogging, air is breathed in through the mouth and bypasses this filter system. The air then passes through the windpipe (*trachea*). The trachea divides in two (bifurcates) to furnish air to each lung through the *bronchi*. Each bronchus divides and redivides about 15 more times; the

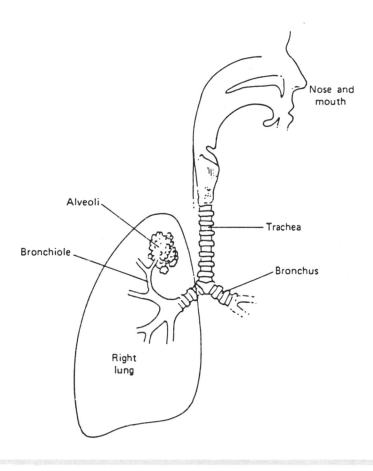

Figure 7.2. A schematic diagram showing the principal air passages into the lungs.

resulting terminal bronchioles supply air to many millions of small sacs called *alveoli*. The alveoli, which are like small interconnected bubbles (Fig. 7.3), are about 0.2 mm in diameter (the page you are reading is about 0.1 mm thick) and have walls only 0.4 μm thick. They expand and contract during breathing; they are "where the action is" in the exchange of O_2 and CO_2. Each alveolus is surrounded by blood so that O_2 can diffuse from the alveolus into the red blood cells and CO_2 can diffuse from the blood into the air in the alveolus. At birth the lungs have about 30 million alveoli; by age 8 the number of alveoli has increased to about 300 million (~100,000/day). Beyond this age the number stays relatively constant, but the alveoli increase in diameter. The alveoli play such an important role in breathing that we will discuss the physics of the alveoli in more detail in Section 7.5.

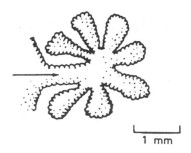

Figure 7.3. The structure of the alveoli.

In addition to serving as the transport system for the air, the airways remove the dust particles that stick to the moist lining of the various air passages. The body has two mechanisms for clearing the airways of foreign particles. Large chunks are removed by coughing. Small particles are carried upward toward the mouth by millions of small hairs, or *cilia*. The cilia, which are only about 0.1 mm long, have a waving motion that moves mucus carrying dust and other small particles up the major airways. Each of the cilia vibrates about 1000 times a minute (frequency $\cong 17$ Hz). The mucus moves at about 1 to 2 cm/min (about 100–200 m/week!). You can think of the cilia as an escalator system for the trachea. It takes about 30 min for a particle of dust to be cleared out of the bronchi and trachea into the throat where it is spit out or swallowed.

7.2 How the Blood and Lungs Interact

The primary purposes of breathing are to bring a fresh supply of O_2 to the blood in the lungs and to dispose of the CO_2. In this section we will explain the physics involved in the exchange of gas between the lungs and the blood.

Blood is pumped from the heart to the lungs under relatively low pressure (see Chapter 8). The average peak blood pressure in the main pulmonary artery carrying blood to the lungs is only about 2.7 kPa (~ 20 mm Hg) or about 15% of the pressure in the main body circulation. The lungs offer little resistance to the flow of blood. On the average, about one fifth (~ 1 liter) of the body's blood supply is in the lungs, but only about 70 ml of that blood is in the capillaries of the lungs at any one time. Since blood is in the pulmonary capillaries for less than

1 s, the lungs must be well designed for gas exchange; the alveoli of the lungs have extremely thin walls and are surrounded by the blood in the pulmonary capillary system. As stated earlier, the surface area between air and blood in the lungs is about 80 m^2. If the 70 ml of blood in the pulmonary capillaries were spread over a surface area of 80 m^2, the resulting layer of blood would be only about 1 μm thick, less than the thickness of a single red blood cell!

Two general processes are involved in gas exchange in the lungs: (1) getting the blood to the pulmonary capillary bed (*perfusion*) and (2) getting the air to the alveolar surfaces (*ventilation*). If either process fails, not all the blood will be properly oxygenated.

There are three types of ventilation-perfusion areas in the lungs: (1) areas with good ventilation and good perfusion, (2) those with good ventilation and poor perfusion, and (3) those with poor ventilation and good perfusion. In a normal lung the first type accounts for over 90% of the total volume. If the blood flow to part of a lung is blocked by a clot (a pulmonary embolism) that volume will have poor perfusion. If air passages in the lungs are obstructed as in pneumonia, the involved area will have poor ventilation. Many pulmonary diseases cause reductions in perfusion or in ventilation.

The transfer of O_2 and CO_2 into and out of the blood is controlled by the physical law of *diffusion* (Chapter 6, *Osmosis and the Kidneys*). All molecules are continually in motion. In gases and liquids, and to a certain extent even in solids, the molecules do not remain in one location. For example, if you could identify a group of molecules in a room (e.g., from a drop of perfume) in a few minutes you would find that these molecules had moved (diffused) throughout the room. Molecules of a particular type diffuse from a region of higher concentration to a region of lower concentration until the concentration is uniform. In the lungs we are concerned with diffusion in both gas and liquids. In the O_2 and CO_2 exchange in the tissues we are concerned only with diffusion in liquids. The molecules in a gas at room temperature move at about the speed of sound. Each molecule collides about 10^{10} times/s with neighboring molecules, in the process wandering about in a random manner. The most probable distance D a molecule will travel from its origin after N collisions is $D = \lambda\sqrt{N}$ where λ is the mean free path, or the average distance between collisions. In air λ is about 10^{-7} m; in tissue λ is about 10^{-11} m (see Problem 7.2).

Diffusion depends on the speed of the molecules; it is more rapid if the molecules are light and it increases with temperature. Since N is proportional to the diffusion time Δt (i.e., $N \propto \Delta t$), we can write that

$D \propto \sqrt{\Delta t}$, or $\Delta t \propto D^2$. If $D = 10$ mm after 1 s, it will take that molecule 100 s on the average to diffuse 100 mm. In the lungs the distance to be traveled is usually a small fraction of a millimeter, and diffusion takes place in a fraction of a second. The diffusion of O_2 and CO_2 in tissue is about 10,000 times slower than it is in air, but the tissue thickness the molecules must diffuse through in the lungs is very small (~ 0.4 μm) and diffusion through the alveolar wall takes place in much less than 1 s. We discuss diffusion in tissues more in Chapter 8.

To understand the behavior of gases in the lungs it is necessary to review Dalton's law of partial pressures. This law says that if you have a mixture of several gases, each gas makes its own contribution to the total pressure as though it were all alone. Consider a closed liter container of dry air at atmospheric pressure 101 kPa (760 mm Hg). If you removed all the molecules except O_2 from the container the pressure would drop to about 20 kPa (20% of 101 kPa, or 150 mm Hg). This is the partial pressure of oxygen, pO_2. If only the N_2 molecules were left, the pressure would be about 80% of 101 kPa, or about 80 kPa (610 mm Hg). Figure 7.4 schematically shows this imaginary experiment. The partial pressure of water vapor in air depends on the humidity. In typical room air the partial pressure of water vapor is 2 to 2.6 kPa (15 to 20 mm Hg); in the lungs at 37 C and 100% relative humidity the partial pressure of water vapor is 6.2 kPa (47 mm Hg).

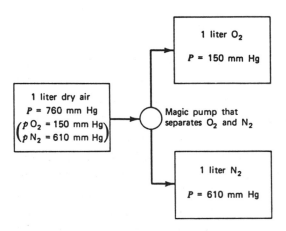

Figure 7.4. A schematic illustration of Dalton's law of partial pressures. A liter of air at 101 kPa (760 mm Hg) pressure can be thought of as a mixture of 1 liter of O_2 at a pressure of 20 kPa (150 mm Hg) and 1 liter of N_2 at a pressure of 80 kPa (610 mm Hg).

The mixture of gases in the alveoli is not the same as the mixture of gases in ordinary air. The lungs are not emptied during expiration. During normal breathing the lungs retain about 30% of their maximum volume at the end of each expiration. This is called the *functional residual capacity* (FRC). At each breath about 0.5 liter of fresh air (pO$_2$ of 20 kPa) mixes with about 2 liters of stale air in the lungs to result in alveolar air with a pO$_2$ of about 13 kPa. The pCO$_2$ in the alveoli is about 5 kPa. Expired air includes about 0.15 liters of relatively fresh air from the trachea that was not in contact with alveolar surfaces, so expired air has a slightly higher pO$_2$ and a lower pCO$_2$ than alveolar air (Table 7.1). The ratio of CO$_2$ output to O$_2$ intake is called the respiratory exchange ratio or respiratory quotient R (see Chapter 2, *Energy, Heat, Work, and Power of the Body*). R is usually slightly less than 1, and depends on the type of food we have eaten recently.

Table 7.1. The Percentages and Partial Pressures of O$_2$ and CO$_2$ in Inspired, Alveolar, and Expired Air[a]

	% O$_2$	pO$_2$ (kPa)	% CO$_2$	pCO$_2$ (kPa)
Inspired air	20.9	20	0.04	0.04
Alveolar air	14.0	13	5.6	5.3
Expired air	16.3	15	4.5	4.3

[a] It is assumed that the inspired air is dry and the expired air is saturated, pH$_2$O = 6.2 kPa.

Consider what happens in a closed container of blood and O$_2$. Some O$_2$ molecules collide with blood and are dissolved. After a while the number of O$_2$ molecules that are escaping from the blood each second is the same as the number that are entering it. The blood then has a pO$_2$ equal to that of the O$_2$ in contact with it. If the pO$_2$ in the gas phase is doubled, the amount of O$_2$ dissolved in the blood will also double. This proportionality is called *Henry's law of solubility of gases*.

The amount of gas dissolved in blood varies greatly from one gas to another. Oxygen is not very soluble in blood or water. At body temperature, 1 liter of blood plasma at a pO$_2$ of 13 kPa (~100 mm Hg) will hold only about 2.5 cm^3 of O$_2$ at normal temperature and pressure (NTP). At a pCO$_2$ of only 5 kPa (~40mm Hg) it will hold about 25 cm^3 of CO$_2$ in each liter. If the body had to depend on dissolved O$_2$ in the plasma to supply O$_2$ to the cells, the heart would have to pump 140 liters of blood per minute at rest instead of the 6 liters/min it actually pumps. As we discuss shortly there is a more efficient method of transporting O_2 and CO_2 that involves the red blood cells.

The different solubilities of O_2 and CO_2 in tissue affect the transport of these gases across the alveolar wall. A molecule of O_2 diffuses faster than a molecule of CO_2 because of its smaller mass. However, because of the greater number of CO_2 molecules in solution, the transport of CO_2 is more efficient than the transport of O_2. If a disease causes the alveolar wall to thicken, the transport of O_2 is hindered more than the transport of CO_2.

Nitrogen from the air does not play any known role in body function, save for the fact that pure O_2 would be toxic; N_2 plays a safety role here. It is dissolved in the blood at its partial pressure. A deep-sea diver breathes air at a much higher pressure underwater than at sea level; as a result, the increased partial pressure of N_2 causes more N_2 to be dissolved in his blood and tissues. If the diver surfaces too rapidly some of the N_2 forms bubbles in the joints causing the serious problem of "bends" (see Chapter 5).

During normal breathing the fresh supply of air does not reach the alveoli which are still filled with stale air from previous breaths. Because of its higher concentration, the fresh O_2 rapidly diffuses through the stale air to reach the surface of the alveoli. The O_2 is dissolved in the moist alveolar wall and diffuses through into the capillary blood until the pO_2 in the blood is equal to that in the alveoli. This process takes less than 0.5 s (Fig. 7.5). Meanwhile the CO_2 in the blood diffuses even more rapidly into the gas in the alveoli until the pCO_2 in the alveolar gas is the same as in the blood.

As mentioned earlier, the blood can carry very little O_2 in solution. Most of the O_2 for the cells is carried in chemical combination with the hemoglobin (Hb) in the red blood cells. Each red blood cell can carry about a million molecules of O_2. A liter of blood can carry about 200 cm^3 of O_2 at NTP by this means while it can carry only 2.5 cm^3 of O_2 in solution. Since most of the O_2 is not in solution, the law of diffusion is altered. The O_2 will combine with or separate from the Hb in a way that depends on the dissociation curve (Fig. 7.6). The Hb leaving the lungs is about 97% saturated with O_2 at a pO_2 of about 13 kPa (100 mm Hg). The pO_2 has to drop by about 50% before the O_2 load of the blood is noticeably reduced (Fig. 7.6).

When the blood reaches the cells with their low pO_2 environment, the O_2 is dissociated from the Hb and diffuses into the cells. Not all the O_2 leaves the Hb; the amount leaving depends on the pO_2 of the

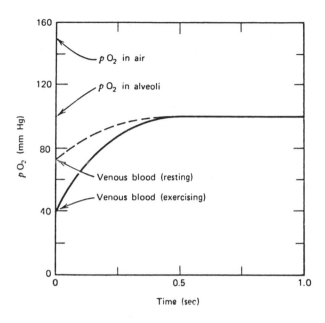

Figure 7.5. The pO_2 of the blood in a pulmonary capillary rises rapidly to the level of the pO_2 in the alveoli as the red blood cells move through the capillary (dashed line). Even during heavy exercise (solid line) the red blood cells are rapidly replenished with O_2.

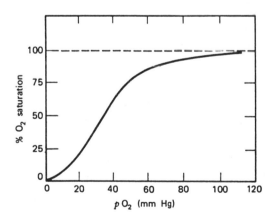

Figure 7.6. The percent O_2 saturation of the blood as a function of pO_2 in the alveoli. At 100% saturation 1 liter of blood can transport 200 ml of O_2 at NTP. This curve is affected by the temperature, the pCO_2, and the pH.

tissues. Under resting conditions the venous blood returns to the heart with about 75% of its load of O_2. The O_2 is retained in the blood because it is not needed in the tissues. During heavy physical labor or exercise the situation in active muscle changes drastically. The pO_2 in the working muscles drops rapidly causing more O_2 to be dissociated from the Hb and to diffuse into the muscles. In addition the body can increase the blood flow to working muscles by a factor of three. Working muscles can consume 10 times more O_2 than they consume at rest. For normal people the limiting factor in exercise is not the amount of blood pumped by the heart per minute (cardiac output) or the amount of O_2 supplied to the blood by the lungs, but the speed at which O_2 is transferred to the working muscles. Because of the vigorous workout of the heart, even at rest, blood in the coronary arteries is depleted of most of their O_2. Perhaps the coronary arteries are under-designed.

The dissociation of O_2 from Hb also depends on the pCO_2, the pH (acidity), and the temperature. During exercise the pCO_2, the acidity, and the temperature in working muscles all increase; these increases all shift the curve of Fig. 7.6 to the right and permit the Hb to give up more of its O_2. All these factors thus increase the O_2 to the working muscles. In the lungs the decrease of pCO_2 due to rapid breathing permits Hb to bind more O_2.

Carbon dioxide is not transported from the tissues by simple diffusion either. Most of the CO_2 remains in the blood after it has left the lungs (pCO_2 ~40 mm Hg). The CO_2 level in the blood is maintained fairly constant by the breathing rate. Excessively rapid breathing (hyperventilation) can lower the pCO_2 in the blood (hypocapnia); this causes mental disturbances and fainting.

In carbon monoxide (CO) poisoning the CO molecules attach very securely to the Hb at places normally used by the O_2. They attach about 250 times more tightly than O_2 and do not easily dissociate into the tissues. In addition to using places normally used to transport O_2, the CO inhibits the release of O_2 from Hb, so even a small amount of CO can seriously reduce the O_2 to the tissues. Cigarette smokers breathe in about 0.25 liter of CO from each pack, and it is also commonly inhaled by people driving in heavy traffic. Carbon monoxide can cause death by starving the tissues of O_2.

Normally the dissolved O_2 in the blood is of no significance, but if a CO victim is placed in a hyperbaric O_2 chamber with an absolute pressure of 3 atm of pure O_2, the pO_2 increases by a factor of 15. The dissolved O_2 in the blood can then supply minimal body needs (see

Chapter 5, *Pressure in the Body*). This therapy cannot be maintained very long because O_2 poisoning can result. Continued use of 1 atm of pure O_2 can cause swelling (edema) of the lung tissues, which reduces O_2 to the blood and ironically results in death from a lack of O_2 (anoxia). Safe levels of pO_2 in "air" are those below 0.5 atm (50 kPa) (refer to Fig. 5.8).

7.2

PROBLEM

Compute the most probable distance, D, an O_2 molecule will travel in air and in tissue after 1 s if the molecule experiences $N = 10^{10}$ collisions/s in air and $N = 10^{12}$ collisions/s in tissue.
[Answer: $D_{air} = 10^{-2}$ m (1 cm); $D_{tissue} = 10^{-5}$ m (0.01 mm)]

7.3

PROBLEM

If there are 3×10^8 alveoli in a lung with a functional residual capacity (FRC) of 2.5 liters, calculate the average volume of an alveolus.
[Answer: 8.3×10^{-12} m^3]

7.4

PROBLEM

Explain qualitatively why the air in the lungs has a pO_2 of only about 13 kPa while the pO_2 in ordinary air is about 20 kPa.

7.5

PROBLEM

If the thickness of the alveolar walls doubled, by what factor would the diffusion time for O_2 to reach the blood change?
[Answer: 4 times as long]

7.3 Measurement of Lung Volumes

A relatively simple instrument, the *spirometer* (Fig. 7.7), is used to measure airflow into and out of the lungs and to record it on a graph of volume versus time. Figure 7.8 shows a typical recording for an adult under various breathing conditions. During normal breathing at rest we inhale about 0.5 liter (500 cm^3) of air with each breath. This is referred to as the *tidal volume at rest*. At both the beginning and end of a normal breath there is considerable reserve. At the end of a normal inspiration it is possible with some effort to further fill your lungs with air. The additional air taken in is called the *inspiratory reserve volume*. Similarly, at the end of a normal expiration you can force more air out of your lungs. This additional expired air is called the *expiratory reserve volume*. The air remaining in the lungs after a normal expiration is called the *functional residual capacity* (FRC). It is this stale air that mixes with the fresh air of the next breath. During heavy exercise, the tidal volume is considerably larger. You have a fair idea of your lung capacity if you have ever blown up a paper sack or an air mattress. If a person breathes in as deeply as possible (*a* in Fig. 7.8) and then exhales as much as possible (*b* in Fig. 7.8), the volume of air exhaled is called the *vital capacity*. However, the lungs will still contain some air—the *residual volume*, which is about one liter for an adult. The residual volume can be determined by having the subject breathe in a known volume of an inert gas such as helium and then measuring the fraction of helium in the expired gas. Since the helium and air will mix thoroughly during a single breath, this dilution technique is quite accurate.

A number of clinical tests can be made with the spirometer. The amount of air breathed in 1 min is called the *respiratory minute volume*. The maximum volume of air that can be breathed in 15 s is called the *maximum voluntary ventilation* and is a useful clinical quantity. The maximum rate of expiration after a maximum inspiration is a useful test for emphysema and other obstructive airway diseases. In some cases the flow rate decreases with increased expiratory effort. A normal person can expire a volume equal to about 70% of the vital capacity in 0.5 s, 85% in 1.0 s, 94% in 2.0 s, and 97% in 3.0 s. Normal peak flow rates are 350 to 500 liters/min. The velocity of the expired air can be impressive; if a person coughs or sneezes hard without covering the mouth, the velocity of the air in the trachea can reach Mach 1—the velocity of sound in air! This high velocity can cause partial collapse of the airways because of the *Bernoulli effect* (i.e., a higher velocity means a lower pressure in the gas).

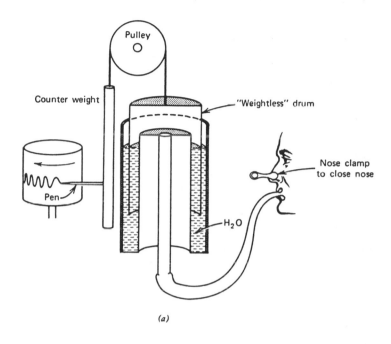

Figure 7.7. The spirometer is used to measure various quantities of pulmonary function. The airflow in and out of the lungs is recorded on a rotating chart. (a) A cross section of a spirometer showing how water is used as an air-tight seal to keep air within the counterbalanced drum. (b) One of the authors (JRC) producing the graph shown in Figure 7.8. The nose clamp forces all air to flow through the mouth.

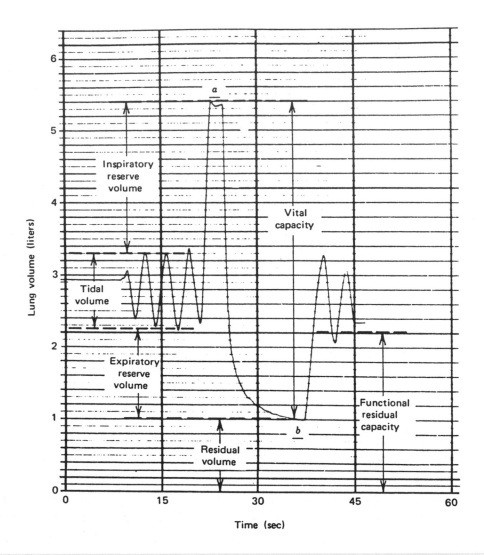

Figure 7.8. A tracing made using the apparatus shown in Figure 7.7b. It shows the various volumes and capacities of the lungs. Note that during the maximum expiration the outflow is rapid at first; the last 5% takes longer than the first 95%.

In coughing to dislodge a foreign object, this partial collapse of the airways increases the air velocity and increases the force on the foreign object. An increase in air pressure in the trachea is also the basis for the well-known Heimlich maneuver which is used to dislodge material from the esophagus when a person is choking. A sudden upward, externally applied impulse on the diaphragm decreases the available lung volume, thus increasing the pressure and (if there is any opening) the resultant velocity of air leaving the lungs.

Not all of the air we inspire adds O_2 to the blood. The volume of the trachea and bronchi is called the *anatomic dead space* since air in this space is not exposed to the blood in the pulmonary capillaries. Typically the anatomic dead space is about 0.15 liters (150 cm^3). In addition, in some diseases some of the alveolar capillaries are not perfused with blood and the O_2 is not absorbed in these alveoli. This unused volume is called *physiologic* or *alveolar dead space*. Air in the dead space does not provide any O_2 to the body. The stale air in the anatomic dead space after an expiration is taken back into the lungs during the next inspiration. If you increase your dead space by breathing through a long tube, you will recycle more of your own breath. If the tube has a volume equal to your vital capacity, you obviously will get no new air and will suffocate.

7.4   Pressure-Airflow-Volume Relationships of the Lungs

The pressure, airflow, and volume relationships of the lungs during tidal breathing for a normal subject and for a patient with a narrowed airway are shown in Fig. 7.9. The pressure difference needed to cause air to flow into or out of the lungs of a healthy individual is quite small. Note that the pressure difference (Fig. 7.9a) is only about 200 Pa (a few centimeters of water) for a normal individual. Figure 7.9b shows the rate of air flow into and out of the lungs in liters/min, and Fig. 7.9c shows the lung volume during the breathing cycle.

Since the esophagus passes through the chest, and normally is closed at both ends, it reflects the pressure between the lungs and chest wall (intrapleural or intrathoracic space). It is possible to measure the pressure in the esophagus with a pressure gauge. This pressure is normally negative (−1.3 kPa or −10 mm Hg) due to the lungs wanting to collapse (see Section 7.6). In Fig. 7.10, the intrathoracic pressure (measured in the esophagus) is plotted versus the tidal lung volume during respiration. Figure 7.11 shows the pressure-volume curves for three different breathing rates—slow, moderate, and fast.

The lungs and chest wall are normally in contact, with the lungs trying to deflate and the chest trying to expand. The behavior of the

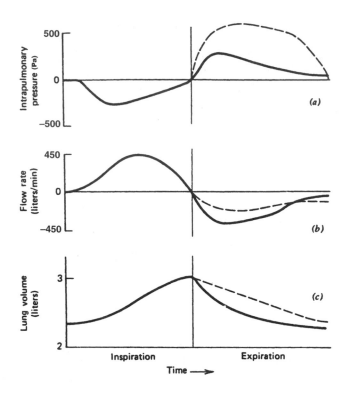

Figure 7.9. Typical pressures (a), flow rates (b), and lung volumes (c) during quiet respiration for a normal individual (solid line) and a patient with a narrowed airway (dashed line). Note the increased pressure and decreased flow rates during expiration in the patient with the narrowed airway.

lungs-chest system is the result of the combination of the physical characteristics of the two. Figure 7.12 shows curves of volume versus pressure for the chest wall and the lungs separately and for the two together. The volume is given as a percentage of the vital capacity. If the chest wall were free of its interaction with the lungs it would expand to a larger volume of about two-thirds of the total vital capacity. The lungs by themselves would collapse and have essentially no air volume. Together the lungs and chest wall come to a relaxation volume (FRC) at about 30% of vital capacity.

The combined curve in Fig. 7.12 shows the pressure-volume (P-V) relationship obtained by filling the lungs to known percentages of the vital capacity. The pressure is measured in the mouth (and lungs) with the nose and mouth closed and the breathing muscles relaxed. For example, at about 60% of vital capacity, the relaxation pressure is 1 kPa (10 cm H_2O). Since the chest wall is at equilibrium at this volume, this pressure is produced by the elastic properties of the lung.

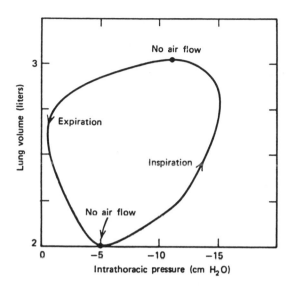

Figure 7.10. Intrathoracic pressure plotted vs. lung volume during respiration for a larger than average tidal volume. (Adapted from Hildebrandt, J., and A. C. Young, in T. C. Ruch and H. D. Patton (Eds.), *Physiology and Biophysics*, 19th ed., © W. B. Saunders Company, Philadelphia, 1965, p. 754.)

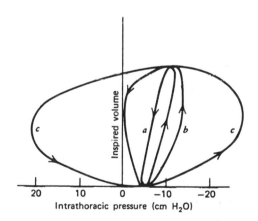

Figure 7.11. The P-V curves for three different breathing rates: (a) very slow breathing of about 3 breaths/min; (b) about 40 breaths/min; and (c) maximum breathing rate of about 150 breaths/min (10 cm H_2O = 1 kPa). (Adapted from Hildebrandt, J., and A. C. Young, in T. C. Ruch and H. D. Patton (Eds.), *Physiology and Biophysics*, 19th ed., © W. B. Saunders Company, Philadelphia, 1965, p. 754.)

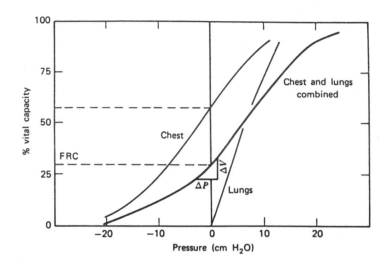

Figure 7.12. The P-V curves for the chest alone, the lungs alone, and the chest and lungs combined. The combined curve is the relaxation curve of Fig. 7.13 (10 cm H₂O = 1 kPa). The slope of the combined curve ΔV/ΔP gives the compliance of the lung-chest system. If the vital capacity is 5 liters, ΔV/ΔP = 2 l/kPa. (Adapted from Hildebrandt, J., and A. C. Young, in T. C. Ruch and H. D. Patton (Eds.), *Physiology and Biophysics*, 19th ed., © W. B. Saunders Company, Philadelphia, 1965, p. 759.)

When the same relaxation measurements are made after a forced exhalation the negative pressure values of Fig. 7.12 are obtained.

The relaxation pressure curve is again plotted as a function of the vital capacity in Fig. 7.13. In addition, two other related curves are shown. All of these pressures are measured in the mouth with the nose and mouth closed. Exhaling with the greatest force gives the maximum expiratory effort curve. Inhaling with maximum effort gives the maximum inspiratory effort curve. Forced expiratory effort after a maximum inspiration (100% of vital capacity) compresses the gas according to Boyle's law, PV = constant. The dashed lines a and b show the theoretical curves for the pressure-volume relationship of an ideal gas (PV = constant) at 0% and 100% of vital capacity.

Compliance is an important physical characteristic of the lungs. Compliance is the ratio of the change in volume produced by a small change in pressure, that is, $\Delta V / \Delta P$ (see Fig. 7.12). Compliance is usually given in liter/cm H₂O. Compliance in normal adults is in the range of 0.18 to 0.27 liter/cm H₂O. It is generally about 25% greater in men over age 60 than in younger men. There is little change in women with age.

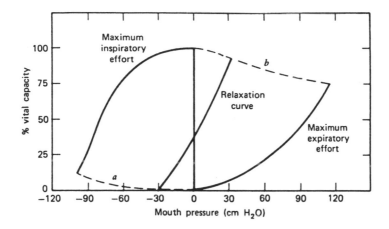

Figure 7.13. P-V curves obtained with a pressure gauge in the mouth. The center curve is for the lung-chest system shown in Fig. 7.12. The curve on the right is the maximum pressure obtained when the subject blows as hard as possible on the gauge. The curve on the left is obtained by maximum suction. The dashed curves a and b are the theoretical curves for Boyle's law PV = constant. (Adapted from Hildebrandt, J., and A. C. Young, in T. C. Ruch and H. D. Patton (Eds.), *Physiology and Biophysics*, 19th ed., © W. B. Saunders Company, Philadelphia, 1965, p. 758 after Rahn et al., *Amer. J. Physiol.*, 146, 1946, pp. 161–178.)

A stiff (*fibrotic*) lung has a small change in volume for a large pressure change and thus it has a low compliance. A flabby lung has a large change in volume for a small change in pressure and has a large compliance. Infants with respiratory distress syndrome (see Section 7.5) have lungs with low compliance. In some diseases, such as emphysema, the compliance increases (see Section 7.9).

During tidal breathing, the P-V curve forms a closed loop like those shown in Fig. 7.14. The cycles flow clockwise on the loops. The middle loop represents typical tidal breathing at normal pressure. Loop *b* represents positive pressure breathing where the air supply pressure is about 25 cm H_2O greater than the pressure on the chest wall. Positive pressure breathing is often used therapeutically in resuscitation and in relief of obstructive airway disease. For positive pressure breathing the inspiratory muscles are not used but the expiratory muscles are. Loop *c* in Fig. 7.14 represents negative pressure breathing. This can occur when a person is underwater and breathing through a tube to the surface (snorkel breathing). In this case the inspiratory muscles never completely relax.

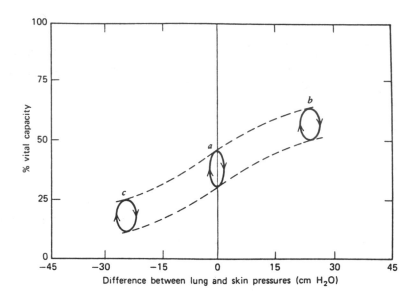

Figure 7.14. P-V curves for tidal breathing under three conditions: (a) normal breathing where the pressure in the mouth is the same as on the skin; (b) positive pressure breathing with a tight-fitting face mask where the breathing muscles must work to expire; and (c) snorkel (underwater) breathing where the pressure on the chest is greater than in the mouth and the inspiratory muscles are under continuous tension (10 cm H_2O = 1 kPa). (Adapted from Hildebrandt, J., and A. C. Young, in T. C. Ruch and H. D. Patton (Eds.), *Physiology and Biophysics*, 19th ed., © W. B. Saunders Company, Philadelphia, 1965, p. 758 after Rahn et al., *Amer. J. Physiol.*, 146, 1946, pp. 161–178.)

7.6

PROBLEM

(a) What are the factors that determine how deep you can swim while breathing through a snorkel to the surface? (b) Estimate the maximum depth at which you could use a snorkel (Hint: see Fig. 7.13).

7.7

PROBLEM

In Morochocha, Peru, the atmospheric pressure is 59 kPa (447 mm Hg). Find the pO_2 and pN_2 that people in this village inhale.
[Answer: pO_2 = 12 kPa (89 mm Hg); pN_2 = 47 kPa (358 mm Hg)]

7.8

PROBLEM

A person's lung volumes were measured and the following results were obtained: vital capacity: 5 liters; residual volume: 1.0 liter; and expiratory reserve volume: 1.5 liters. Find the functional residual capacity (FRC) of this individual. [Hint: See Figure 7.8]
[Answer: 2.5 liters]

7.9

PROBLEM

A person inspired maximally and then began breathing from an expandable bag containing 2 liters of 40% helium gas. After a few breaths, the helium concentration in the bag was 10%. What was this person's total lung capacity?
[Answer: 6 liters]

7.5 Physics of the Alveoli

The alveoli are physically like millions of small interconnected bubbles. They have a natural tendency to get smaller due to the surface tension of their unique fluid lining. This lining, a type of surfactant, is necessary for the lung to function properly. The absence of surfactant in the lungs of some newborn infants, especially prematures, is the cause of the idiopathic respiratory distress syndrome (RDS), sometimes called hyaline membrane disease. (Idiopathic means that its cause is unknown.) This disease causes many infant deaths each year in the United States.

To understand the physics of the alveoli we have to understand the physics of bubbles. The pressure inside a bubble is inversely proportional to the radius and directly proportional to the surface tension γ (gamma). The exact relation is $P = 4\gamma/R$, a form of Laplace's law. Consider soap bubbles on the ends of a tube with a valve separating them as shown in Fig. 7.15a. What happens when the valve is opened to connect them? Because the smaller one has a higher internal pressure (R is small in the equation), it will empty its air into the larger one until the radii of curvature of the large bubble and of the remainder of the small bubble are the same (Fig. 7.15b). Although alveoli are not exactly the same as soap bubbles, there is a tendency for the smaller alveoli to collapse. The condition that results when a sizable number collapse is

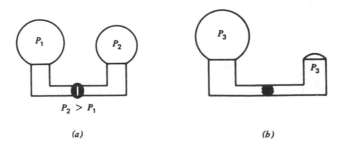

Figure 7.15. The pressure inside a soap bubble depends on its radius. (a) When the valve between the two bubbles is closed, the pressure is greater in the smaller bubble (P = 4γ/R). (b) When the valve is opened, the smaller bubble empties into the larger, leaving the spherical cap with the same radius as the new bubble.

called *atelectasis*. The reason most alveoli do not collapse is related to the unique surface tension properties of surfactant.

The surface tension, γ, of a fluid can be found by measuring how much force is necessary to pull a loop of wire from a clean liquid surface (Fig. 7.16). The surface tension of a water-air interface is 72×10^{-5} N/m; that of a plasma-air interface is about 40 to 50×10^{-5} N/m; those of detergent solutions in air are from 25 to 45×10^{-5} N/m. A qualitative measure of surface tension is to note how long small bubbles of a liquid survive. The lower the surface tension, the longer they last. Observations have shown that bubbles expressed from the lung are very

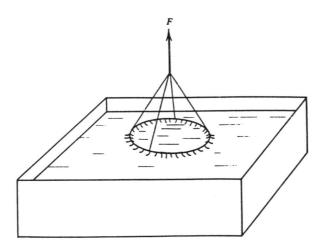

Figure 7.16. The surface tension of a liquid can be measured by determining the force needed to pull a wire loop from a clean liquid surface.

stable, lasting for hours. It can be concluded from this that they must have a very low surface tension, and thus low pressure in the bubble.

The surface tension of the surfactant that lines the alveoli of healthy individuals plays a major role in lung function. The surface tension of the surfactant is not constant. Figure 7.17b shows the surface tension of a film of normal lung extract containing surfactant. Note the large decrease of γ as the area decreases. This characteristic

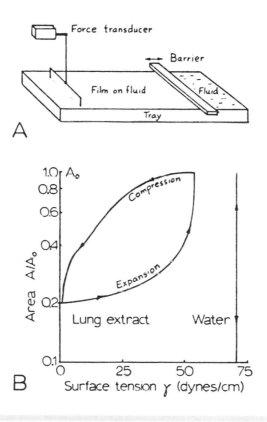

Figure 7.17. Surface tension as a function of film area. (A) A schematic representation of the apparatus used to measure the surface tension of a film. The tray is filled to the top with fluid, a film of the material to be studied is spread on the surface, a movable barrier is used to compress the film, and a hanging plate balance continuously records the surface tension γ. (B) Graph of surface tension of a lung extract containing surfactant. Note the large decrease in surface tension as the area decreases and the different curve obtained as the area increases. The vertical line at about 70 dynes/cm shows that the surface tension of water is constant with changes in the area. Note that 10^3 dyne/cm = 1 N/m. (From Hildebrandt, J., and A. C. Young, in T. C. Ruch and H. D. Patton (Eds.), Physiology and Biophysics, 19th ed., © W. B. Saunders Company, Philadelphia, 1965, p. 744.)

causes the surface tension of the alveoli to decrease as the alveoli decrease in size during expiration. For each alveolus there is a size at which the surface tension decreases sufficiently fast that the pressure starts to drop instead of continuing to increase, and this causes the alveolus to stabilize at about one-fourth its maximum size. Alveoli not covered with surfactant, such as those of infants with RDS, collapse like small bubbles, and quite a large pressure is needed to reopen them. An infant with RDS may not have the energy to breathe with its low compliance lungs. Therapy may involve positive pressure breathing to help open the alveoli.

The P-V curves for an excised human lung are shown in Fig. 7.18. If the lung is completely collapsed, a considerable pressure is needed to start its inflation, similar to the extra effort needed to start blowing up a rubber balloon. From this point the lung inflates rather easily until it is close to its maximum size. The pressure curve during deflation looks quite different. When the pressure has dropped to zero the lung still retains some air. Much less pressure is needed to then reinflate the lung, although reinflation will not follow the deflation curve. A cyclical process in which different curves are followed on the two halves of the

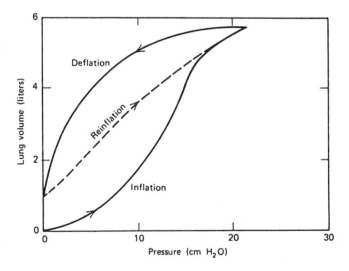

Figure 7.18. Typical P-V curve for an excised human lung. More pressure is required for the initial inflation (lower curve) than for reinflation (dashed curve).

cycle is said to show *hysteresis*. The area inside the loop is proportional to the energy lost as heat during the cycle.

During normal tidal breathing the hysteresis loop is quite small, like curve *a* on the normal P-V curve in Fig. 7.19. If tidal breathing continues unchanged, some of the alveoli collapse, and the hysteresis loop becomes slightly larger and shifts toward higher pressure as shown by curve *b*. A deep breath reopens the alveoli, and the curve shifts back to *a*. We take such a deep breath occasionally without being

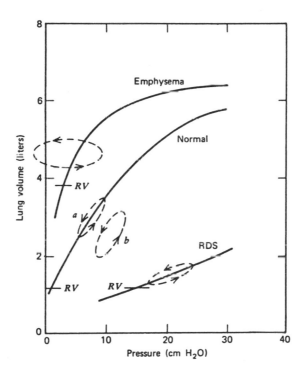

Figure 7.19. P-V curves for a normal subject and for patients with emphysema and RDS. The solid lines are static values. The pressure is relative to the intrathoracic pressure. RV is the residual volume. Notice the increased FRC and RV for emphysema. The dashed lines represent the hysteresis loops during tidal breathing. Curve *a* gradually shifts to become curve *b* as atelectasis and surfactant changes occur. A deep breath shifts curve *b* back to curve *a* as the collapsed alveoli are forced open again. (Adapted from Hildebrandt, J., and A. C. Young, in T. C. Ruch and H. D. Patton (Eds.), *Physiology and Biophysics*, 19th ed., © W. B. Saunders Company, Philadelphia, 1965, pp. 740–758.)

aware of it—a sigh. During surgery the anesthesiologist will occasionally force a large volume of gas into the patient's lungs to reopen collapsed alveoli. Taping the chest, such as for a rib injury, prevents a patient from taking a deep breath and some of the lung space is likely to be lost by collapsed alveoli, or atelectasis.

The P-V curve for an infant lung suffering from RDS is also shown in Fig. 7.19; the hysteresis loop is shifted to the right and a large pressure must be maintained to keep the lung inflated. Note the lower compliance ($\Delta V/\Delta P$) of this lung. The P-V curve of a patient with severe emphysema is also shown in Fig. 7.19; note the increased compliance, the larger residual volume, and the large area inside the hysteresis loop.

7.6 The Breathing Mechanism

Breathing is normally under unconscious control. Although your rate of breathing can be changed at will, you are usually unaware of your breathing most of the time unless you suffer from asthma or emphysema. The physiological control of breathing depends on many factors, but the pH in the respiratory center of the brain exerts primary control.

If a lung were removed from the chest, all the air would be squeezed from it and it would collapse to about one-third of its size—much as a balloon collapses when air is let out of it. The lung can be thought of as millions of small balloons, all trying to collapse. Refer to the demonstration in Fig. 7.20. The lungs do not normally collapse because they are in an airtight container—the chest. As the diaphragm and rib cage move, the lungs stay in contact with them. Two forces keep the lungs from collapsing: (1) surface tension between the lungs and the chest wall and (2) air pressure inside the lungs. The surface tension force is similar to that between two pieces of cellophane or plastic wrap stuck together. If the lungs overcame this force and pulled away from the chest wall, a vacuum would be created since air cannot reach the intrapleural space. Since the air inside the lungs is at atmospheric pressure—100 kPa (14.5 lb/in.^{2}) it would push the lungs back in contact with the chest wall. There is normally a negative pressure of 6 to 12 kPa (5 to 10 mm Hg) in the intrapleural space.

Various muscles are involved in breathing. Intercostal muscles in the chest wall cause the chest to expand when they contract. Normally

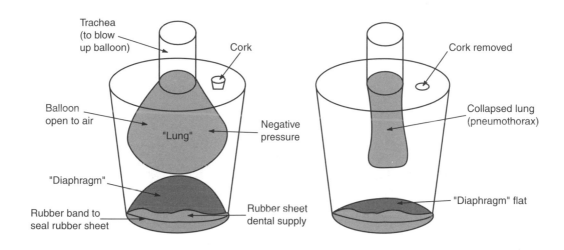

Figure 7.20. The physics of the lungs can be demonstrated crudely by a balloon being sealed into a transparent plastic container (representing the chest) with a rubber sheet (available from a dental supply house) firmly sealed over the large open end (representing the diaphragm). The balloon neck is an airtight seal in the top of the "chest." In order to blow up the balloon, an exit hole is needed—the cork is removed from the small hole in the top. (a) After the balloon is inflated, the cork is replaced. As the air starts to exit from the balloon, a negative pressure is produced in the chest cavity causing the "diaphragm" to be sucked into a dome shape. If the elasticity of the balloon and of the rubber "diaphragm" are appropriately matched, the balloon ("lung") will retain some air. If one pulls down on the "diaphragm," you can see air enter the "lung." (It is difficult to get hold of the "diaphragm" to pull down.) (b) A "pneumothorax"—air in the chest cavity—can be simulated by removing the cork to let air enter the "chest" and the "lung" collapses. In this crude model the diaphragm also flattens. Each lung is in its own compartment and normally both lungs do not collapse at the same time.

most breathing is done by contracting the diaphragm muscles; these pull the diaphragm down, expanding the lungs. When we inspire, we pull the diaphragm down as shown schematically by the arrow in Fig. 7.21b. This produces a slight negative pressure in the lungs and air flows in. When we expire, we relax the diaphragm muscles, the elastic forces in the lungs cause the diaphragm to return to its neutral position, and air flows out of the lungs without any active muscular effort. If the diaphragm muscles are paralyzed, the muscles in the chest wall are used for breathing.

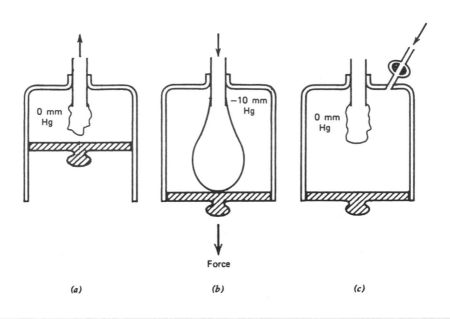

Figure 7.21. A simple model of the mechanism of breathing (a) during expiration, (b) during inspiration, and (c) during pneumothorax.

If the chest wall is punctured as shown schematically by the open valve in Fig. 7.21c, the lung collapses, the diaphragm lowers, and the chest wall expands. This condition is known as a *pneumothorax* (literally, air-chest). Occasionally, such as in the treatment of tuberculosis, it is medically desirable to collapse one lung to allow it to "rest." Since each lung is in its own sealed compartment, it is possible to collapse one lung only as shown in Fig.7.22. This is done rather simply by inserting a hollow needle between the ribs (an intercostal puncture) and allowing air to flow into the intrathoracic space. The air trapped in the space is gradually absorbed by the tissues, and the lung expands to normal over a period of a few weeks. Sometimes a lung collapses spontaneously with no known cause. This condition of *spontaneous pneumothorax* is moderately common in college-age students. As in the medical procedure, the lung returns to normal as the air is absorbed into the surrounding tissues. The chest pains of a collapsed lung are similar to the pains from a heart attack. It is a much less serious condition.

Since both the lung and chest wall are elastic, we can represent them with springs (Fig. 7.23). Under normal conditions they are coupled together: the "lung" springs are stretched and the "chest" springs are compressed (Fig. 7.23a). During a pneumothorax, the lungs and chest are independent and

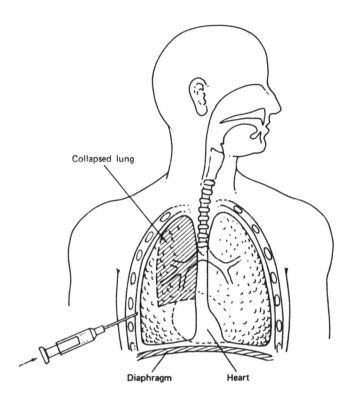

Figure 7.22. A right pneumothorax is produced by letting in air between the chest wall and the lung. The shaded area shows the outline of the collapsed lung.

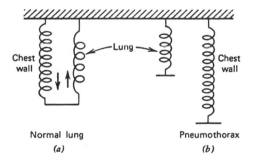

Figure 7.23. (a) A spring model for the lungs. The arrows show the direction of the spring forces. Normally the lung and chest wall are coupled together. (b) During a severe pneumothorax the springs go to their relaxed positions—the chest enlarges and the lung collapses.

the springs representing them go to their relaxed positions as indicated in Fig. 7.23b. The lung collapses, and the chest wall enlarges.

The intrathoracic space is not always at negative pressure. If you close your windpipe and forcefully try to expire, the intrathoracic pressure can become quite high. This is called a *Valsalva* maneuver. A person does this when blowing up a stiff balloon. Under more physiological conditions this is done just before coughing or sneezing and during the stress of defecation or vomiting. Increasing the pressure in the chest compresses the main vein (vena cava) carrying blood back to the right heart and reduces the volume of blood being pumped to the lung. The normal negative pressure in the chest helps keep the vena cava open. The venous blood pressure in the vena cava is only 50 Pa (0.5 cm H_2O) near the heart.

7.7 Airway Resistance

We can breathe in more rapidly than we can breathe out. During inspiration the forces on the airways tend to open them further; during expiration the forces tend to close the airways and thus restrict airflow. For a given lung volume, the expiratory flow rate reaches a maximum and remains constant; it might even decrease slightly with increased expiratory force. Patients with obstructive airway disease such as asthma or emphysema find that an increased effort to breathe out decreases the flow rate considerably. These patients unconsciously find some relief by retaining a large amount of air in the lungs, thus keeping their airways as large as possible. They can often inspire at near normal rates so they breathe in rapidly to allow more time for expiration. The pulmonary physics of emphysema is discussed in Section 7.9.

The flow of air in the lungs is analogous to the flow of current in an electrical circuit. "Ohm's law" for airflow looks like Ohm's law for electrical circuits, with voltage replaced by pressure difference ΔP and current replaced by the volume rate of airflow, $\Delta V/\Delta t$. Airway resistance R_g is the ratio of ΔP to $\Delta V/\Delta t$. Airway resistance is given in units of pressure per unit flow rate, commonly stated in Pa/liter/s or (cm H_2O/liter/s). In typical adults $R_g = 330$ Pa/liter/s (3.3 cm H_2O/liter/s). R_g depends on the dimensions of the tube and the viscosity of the gas. The situation is complicated by the complexity of the airways. Most of the resistance is in the upper airway passages. The nasal area accounts for about half of R_g, and another 20% is due to the other upper airways. In normal subjects less than 10% of R_g is in the terminal airways. Thus diseases that affect the terminal airways (bronchioles and alveoli) do not appreciably affect the airway resistance until they are far advanced.

> The upper airway resistance can be demonstrated with a little practice using a clear drinking straw held firmly between the lips and immersed in a colored fluid, such as tomato juice. When inhaling or exhaling the pressure in the mouth differs from atmospheric pressure due to airway resistance between the back of the throat and the nostrils. If one breathes through the nose, the pressure in the mouth will decrease slightly when inhaling, causing the fluid to rise in the straw. On exhaling, the pressure in the back of the throat is slightly greater than atmospheric, causing the fluid to depress. Breathing more rapidly produces larger excursions in the fluid level of the straw. It takes some practice to breathe through the nose since there is a tendency to suck fluid through the straw.

The time constant of the lungs is related to the airway resistance R_g, and the compliance C. Remember that compliance is $\Delta V/\Delta P$ (Section 7.4). The product R_gC is the time constant for the lung. This is analogous to the time constant RC' for a capacitance C' to discharge through a resistance R in an electrical circuit. (Note that here we use C for compliance and C' for capacitance.) The time constant of the lung is complicated since many parts of the lung are interconnected. If one part of the lung has a larger time constant than other parts, it will not get its share of the air and that part of the lung will be poorly ventilated.

7.10

PROBLEM

The compliance of a normal adult lung is about 0.2 liter/cm H_2O. What is the compliance in m^3/Pa?
[Answer: Approximately 2×10^{-6} m^3/Pa]

7.11

PROBLEM

If $R_g = 300$ Pa(liter/s), what air volume flow rate, V/s, would occur at an expiratory pressure of 13 kPa?
[Answer: 43 liters/s]

7.8 The Work of Breathing

The amount of work done in normal breathing accounts for a small fraction of the total energy consumed by the body (~2% at rest). The primary work of breathing can be thought of as the work done in stretching the springs representing the lung-chest wall-diaphragm system, Fig. 7.24a, which is proportional to the shaded area in Fig. 7.24b; however, this is an oversimplification of the work of breathing. A better model is shown in Fig. 7.25. The resistance of the tissues and the resistance of the gas flow produce heat; these can be represented as a dashpot (R), a resistive element which dampens the motion. The springiness of the lung-chest system is represented by the spring C. The inertia of the mass of the lungs and chest wall must also be overcome; at normal breathing rates, the inertia can be neglected, but at maximum breathing rates (over 100 breaths/min) it is a significant factor. The work of breathing is shown by the total shaded area in Fig. 7.25b; the darker shaded area represents the work against the spring C and the lighter area represents the work against the resistance. During normal breathing, no work is done during expiration; the muscles relax and the springs "snap back" to expel the air, dissipating the energy in the dashpot R. During strenuous exercise muscles are used to expel air. The work of breathing during heavy exercise may amount to 25% of the body's total energy consumption.

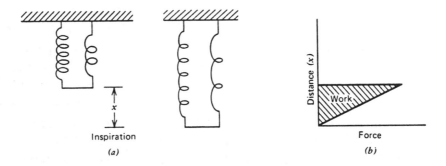

Figure 7.24. A model of the work of breathing. (a) Position of the springs at ends of breathing cycles. (b) The shaded area represents the work done in stretching the springs a distance x.

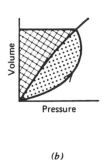

(a)   (b)

Figure 7.25. (a) A better model of the work of breathing. *C* represents the springs of the lung-chest-diaphragm system, *R* is the resistance to tissue motion and gas flow, *I* is the inertia of the moving parts, *P* is the pressure, and *M* represents the breathing muscles. (b) Work done. The cross-hatched shaded area represents the work against the spring *C*, and the dotted area represents the work against the resistance *R*.

Rapid shallow breathing and slow deep breathing are both less efficient than the normal rate. Most animals adjust their breathing rates at rest to use minimum power. At low breathing rates most of the work is done against the elastic forces of the lung and chest (darker area in Fig. 7.25b); at fast breathing rates the work against the resistive forces (lighter area in Fig. 7.25b) increases.

Another way to determine the work done in breathing is to measure the extra O_2 consumed as the breathing rate is increased under resting conditions. The amount of O_2 consumed is directly related to the calories of food "burned" (Chapter 2, *Energy, Heat, Work, and Power of the Body*). We assume that the additional O_2 is used in the respiratory muscles. Figure 7.26 shows a typical curve for a normal subject and the curve for a patient with severe emphysema. The latter may use more O_2 in the work of breathing at a faster rate than is provided by his increased ventilation; the amount of O_2 in his general circulation thus falls.

If we compare the energy used in breathing obtained by the O_2 consumption method to the calculated work done using the model shown in Fig. 7.25, we can estimate the efficiency of the breathing mechanism. Because of many uncertainties, the efficiency estimates range from 5 to 10%.

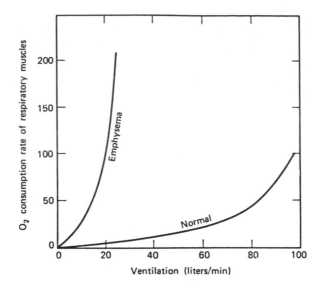

Figure 7.26. Curves of O_2 consumption rate of the respiratory muscles for a normal subject and for a subject with severe emphysema. (Adapted from E. J. M. Campbell, E. K. Westlake, and R. M. Cherniak, *J. Appl. Physiol.* 11:303–308, 1957, Fig. 3.)

7.9 Physics of Some Common Lung Diseases

Diseases of the lungs account for a large percentage of all medical problems. It is estimated that 15% of the people in the United States over age 40 have detectable lung disease. Many of these diseases can be understood in terms of physical changes in the lungs. This does not, of course, mean that a physicist can cure them. The physical aspects of some common lung diseases are discussed in this section. The physics of RDS in infants was discussed in Section 7.5.

At rest, only a small fraction of the lungs' capacity is used. Thus a lung disease that reduces the capacity often does not produce noticeable symptoms in its early stages. When the symptoms are noticeable, the disease is fairly well advanced. Many lung function tests force the breathing mechanism to its limits and thus allow detection of changes that are not ordinarily apparent. There are some simple lung tests that should be included in every health checkup.

In emphysema the divisions between the alveoli break down, producing larger lung spaces. This destruction of lung tissue reduces the springiness of the lungs. The lungs become more compliant—a small change in pressure produces a larger than normal change in the volume. While at first glance this would appear to make it easier to breathe, the opposite is true. Much of the work of breathing is done in overcoming the resistance of the airways. In emphysema the airway resistance increases greatly.

Figure 7.27 will help you understand the physics of emphysema. You can think of the elasticity of the tissues in the normal lung as millions of little interconnected springs (Fig. 7.27a). These "springs" tend to collapse the lung and produce the force that pulls on the chest wall. They also pull on the walls of the airways; this keeps the airways open and helps reduce airway resistance during expiration.

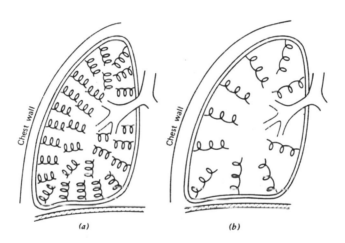

Figure 7.27. Spring models of (a) a normal lung and (b) a lung with severe emphysema. Note the reduced number and strength of the springs in the model for emphysema. The resulting expansion of the chest wall and narrowing of the major airways causes an increase in airway resistance.

The situation in severe emphysema is shown in Fig. 7.27b. The number of working "springs" has been greatly reduced, and those present are much weaker than normal. This produces two important changes: (1) the lung becomes flabby and expands as the reduced tension allows the chest wall to expand almost to the resting volume of the chest wall without the lung—about 60% of vital capacity (Fig. 7.12); and (2) the tissues do not pull very hard on the airways, permitting the narrowed airways to collapse easily during expiration. This increased airway resistance is the major symptom of severe emphysema. The increased size of the lungs increases the FRC and the residual volume (Fig. 7.19). The chest is overinflated, and the posture is affected: someone with the disease appears barrel-chested. Since a person who has emphysema is unable to blow out a candle, it is simple to test for the disease.

Emphysema occurs occasionally in nonsmokers, but the recent large increase in the disease has been primarily among heavy smokers.

In asthma, another common obstructive disease, the basic problem is also expiratory difficulty due to increased airway resistance. Some of this resistance is apparently due to swelling (edema) and mucus in the smaller airways, but much of it is due to contraction of the smooth muscle around the large airways. Lung compliance is essentially normal, but the FRC may be higher than normal since the patient often starts to inspire before completing a normal expiration.

In fibrosis of the lungs the membranes between the alveoli thicken. This has two marked effects: (1) the compliance of the lung decreases, and (2) the diffusion of O_2 into the pulmonary capillaries decreases. The expiratory resistance is essentially normal. A person with the disease will have labored and even painful breathing (dyspnea) or shortness of breath during exercise. Fibrosis of the lungs can occur if the lungs have been irradiated (e.g., in the treatment of cancer), although this is not the only cause.

PROBLEM 7.12 In a patient with severe emphysema, which of the following are above normal and which are below normal?
(a) Airway resistance
(b) Inspiratory reserve volume
(c) Functional residual capacity
(d) Vital capacity
(e) Compliance

8

Physics of the Cardiovascular System

The blood and its supply of O_2 are so important to the body that the heart is the first major organ to develop in the embryo. Eight weeks after conception the heart is working to circulate blood to the tissues of the fetus. Since the fetus does not have functioning lungs, and no way to get air anyway, it must obtain its oxygenated blood from its mother via the umbilical cord. The fetal heart has an opening that permits blood to flow from the right atrium to the left atrium. As a result, only about 10% of the blood is circulated to the fetal lungs. After birth the opening between the right and left atria effectively closes to send much more blood to the lungs. It may take months for the closure to be complete. If closure is not adequate at birth, the blood will not be properly oxygenated and the infant will be a "blue baby." A congenital heart defect of this type can now be corrected with surgery.

The cells of the body act like individual engines. In order for them to function they must have (1) fuel from our food to supply energy, (2) O_2 from the air we breathe to combine with molecules from food to release energy, and (3) a way to dispose of the by-products of the combustion (mostly CO_2, H_2O, and heat). Since the body has about a trillion cells, an elaborate transportation system is needed to deliver the fuel and O_2 to the cells and remove the by-products. The blood performs this important body function. Blood represents about 7% of the body mass or about 4.5 kg (volume ~4.4 liters) in a 64 kg (141 lb) person. The blood, blood vessels, and heart make up the cardiovascular system (CVS). This chapter describes the physical aspects of the CVS.

Several medical specialists are concerned with the CVS. Some physicians who have specialized in internal medicine subspecialize in problems of the blood. They are called hematologists, and they treat patients with blood conditions such as anemia. Other medical specialists called cardiologists are primarily concerned with the heart. Cardiologists treat patients who have had heart attacks, and they interpret electrocardiograms. There is also a subspecialty in surgery that deals exclusively with the CVS. Heart surgeons perform heart-transplant operations, do by-pass procedures, etc. In other medical fields such as radiology and pediatrics there are subspecialists who deal primarily with the CVS.

8.1   Major Components of the Cardiovascular System

Two cross sections of the heart are shown in Fig. 8.1. Basically a double pump, the heart provides the force needed to circulate the blood through the two major circulatory systems: the pulmonary circulation in the lungs and the systemic circulation in the rest of the body (Fig. 8.2). The blood in a normal individual circulates through one system before being pumped by the other section of the heart to the second system.

Let us start with the blood in the left side of the heart and follow its circulation through one complete loop. The blood is pumped by the contraction of the heart muscles from the left ventricle at a pressure of about 17 kPa (125 mm Hg) into a system of arteries that subdivide into smaller and smaller arteries (arterioles) and finally into a very fine meshwork of vessels called the capillary bed. During the few seconds it is in the capillary bed, the blood supplies O_2 to the cells and picks up CO_2 from the cells. After passing through the capillary bed the blood collects in small veins (venules) that gradually combine into larger and larger veins before it enters the right side of the heart via two main veins—the superior vena cava and the inferior vena cava. The returning blood is momentarily stored in the reservoir (the right atrium), and during a weak contraction at a pressure of about 0.8 kPa (5 to 6 mm

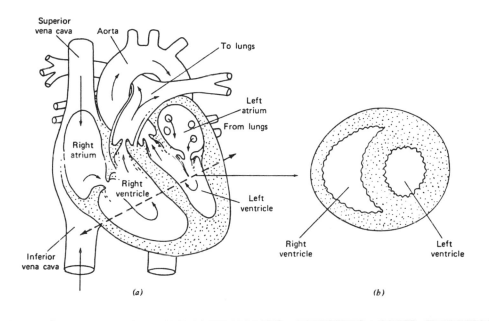

Figure 8.1. The heart. (a) Note the heavier and stronger muscular walls on the left side where most of the work is done. (b) The cross section shows the circular shape of the left ventricle; this shape efficiently produces the high pressure needed for the general circulation.

Hg) the blood flows into the right ventricle. On the next ventricular contraction this blood is pumped at a pressure of about 3.3 kPa (25 mm Hg) via the pulmonary arteries to the capillary system in the lungs. In the lungs the blood receives more O_2 and some of the CO_2 diffuses into the air in the lungs to be exhaled. The freshly oxygenated blood then travels via the main veins from the lungs into the left reservoir of the heart (left atrium); during the weak atrial contraction at a pressure of about 1 kPa (7 to 8 mm Hg) the blood flows into the left ventricle. On the next ventricular contraction this blood is again pumped from the left side of the heart into the general circulation. Since a typical adult has about 4.5 liters of blood and each section of the heart pumps about 80 ml on each contraction, about one minute is needed for the average red blood cell to make one complete cycle of the body.

The heart has a system of valves that, if functioning properly, permit the blood to flow only in the correct direction. If these valves become diseased and do not open or close properly, the pumping of the blood becomes inefficient. Modern developments have made it possible to replace diseased or defective heart valves with artificial mechanical valves or with valves taken from pig hearts, which are very similar to

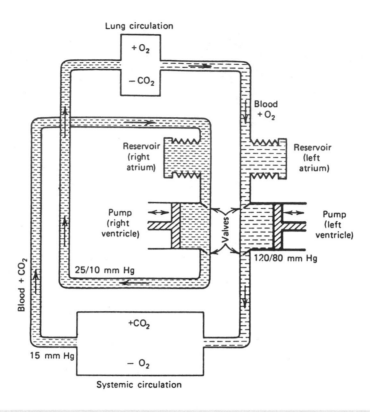

Figure 8.2. The circulatory system can be thought of as a closed-loop circulation system with two pumps. One-way valves keep the flow downward through the pumps. The pressures in mm Hg are indicated.

human valves. It is also possible to implant a heart valve from a cadaver. Both pig valves and cadaver valves must be sterilized by a large amount of ionizing radiation before being implanted.

The blood volume is not uniformly divided between the pulmonary and systemic circulations. At any one time about 80% of the blood is in the systemic circulation and 20% is in the pulmonary circulation. Of the blood in the systemic circulation, about 15% is in the arteries, 10% is in the capillaries, and 75% is in the veins. In the pulmonary circulation, about 7% of the blood is in the pulmonary capillaries and the remaining 93% is almost equally divided between the pulmonary arteries and pulmonary veins.

While we normally think of blood as bright red, most of the blood in the body is dark red. The venous blood is depleted of the O_2 that makes the blood bright red. The blue tint to the veins in your hands is due to pigmentation in the skin. When you cut yourself venous blood usually flows out, as the veins are closer to the surface, but in a fraction of a second it becomes oxygenated and appears bright red.

To the eye, blood appears to be a red liquid slightly thicker than water. When examined by various physical techniques it is found to consist of several different components. The red color is caused by the red blood cells (*erythrocytes*), flat disks about 7 μm (7×10^{-6} m) in diameter, which represent about 45% of the volume of the blood. There are about 5×10^6 erythrocytes/mm^3 of blood. Their typical life span is 3 months; ~10 billion die each day! A nearly clear fluid called blood plasma accounts for the other 55%. The combination of erythrocytes and plasma causes blood to have flow properties different from those of a fluid like water.

Besides erythrocytes and plasma, there are some important blood components, such as the white blood cells (*leukocytes*), present in small amounts. Leukocytes, which are not round, have dimensions of 9 to 15 μm in diameter. They are part of the immune system and play an important role in combating disease. There are about 8000 leukocytes/mm^3 of blood. When there is an infection in the body, the number of white blood cells (white count) increases. (In one type of blood cancer, leukemia, there is an excessive production of leukocytes.) Different types of leukocytes respond differently to infection, and physicians commonly ask for a differential count, that is, a count of the different types of leukocytes.

The blood also contains *platelets*. Platelets (1 to 4 μm in diameter) are involved in the clotting function of blood. There are about 3×10^5 platelets/mm^3 of blood. Platelets live only 3 days. This means that about 5 million platelets die each second and an equal number are produced.

The blood acts as the transport mechanism for small amounts of hormones that control chemical processes in the body. Certain electrolytes (metal ions) in the blood are crucial to the proper functioning of the body. For example, 100 ml of blood normally contains about 10 mg of calcium. If the amount of calcium in the blood drops below 4 to 8 mg/100 ml, the nervous system cannot function normally and death by tetany (muscle spasm) can result.

In the past, a blood cell count was usually done by diluting the blood by a known amount, putting a drop on a glass slide under a microscope, and counting the cells. Since this method is very tedious and the accuracy is only about 15%, an easier and more accurate method was sought and developed. The instrument now routinely used in large clinical laboratories for red blood cell counts is the *Coulter counter*. It was invented by Wallace H. Coulter in the 1950s. The principle of operation is shown in Fig. 8.3. The diluted blood is drawn through a small capillary; the cells essentially go through the capillary one at a time; and as they do, they pass between two electrodes that measure the electrical resistance across the capillary. Each red blood cell causes the resistance to change momentarily as it passes. The change in resistance appears as an electrical pulse that is counted in an electronic circuit. Coulter counters have removed much tedium from red blood cell counts (done by medical technologists) and at the same time have improved the accuracy of the measurement.

Unfortunately, Coulter counters cannot distinguish the different types of white blood cells, so differential counts must still be done with a microscope.

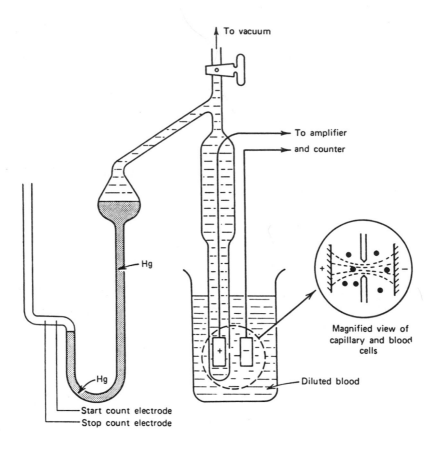

Figure 8.3. A Coulter counter automatically counts blood cells that have been diluted in a conducting solution. The elevated column of Hg produces a reduced pressure and draws the solution through the capillary. As a blood cell passes through the small opening it momentarily increases the resistance between the electrodes. The amplified pulses from the blood cells are counted from the time the Hg touches the start count electrode until the Hg touches the stop count electrode; thus the blood cells in a fixed volume of solution are counted. The insert is a magnified view of the capillary; the electrical current paths are shown as dashed lines.

PROBLEM 8.1

Estimate your blood volume from your mass.

PROBLEM 8.2

Estimate the volume of blood your heart pumps to your systemic circulation each day.

PROBLEM 8.3

What percentage of your blood is in your pulmonary circulation?

PROBLEM 8.4

If leukocytes (white blood cells) have an average diameter of 12 μm, what percentage of the blood volume consists of leukocytes? [Answer: 0.7%]

PROBLEM 8.5

If a platelet has a diameter of about 2 μm, what percentage of the blood volume consists of platelets? [Answer: 0.13%]

(a) About how many erythrocytes (red blood cells) die each minute in an adult? (b) About how many platelets die each minute in an adult?

8.2 O_2 and CO_2 Exchange in the Capillary System

In Chapter 7 we discussed the role of diffusion in the lungs. Oxygen and carbon dioxide also diffuse through tissue. The most probable distance, D, that a molecule will travel after N collisions with other molecules with an average distance λ between collisions is $D = \lambda\sqrt{N}$. In tissue the density of molecules is about 1000 times greater than in air; therefore, λ is much longer in air than in tissue. A typical value for λ in water, which can serve as a model for tissue, is about 10^{-11} m, and a molecule makes about 10^{12} collisions/s. Thus after one second in water the most probable diffusion distance is about 10^{-5} m—the diameter of a typical cell—or about a factor of 10^3 less than in air. This very short diffusion distance is the primary reason that the capillaries in tissue must be very close together. In active muscle approximately one-twelfth of the volume is occupied by capillaries; in heart muscle nearly every cell is in contact with a capillary.

If you cut through a piece of active muscle and count the capillaries you will find about 190/mm^2. The average diameter of the capillaries is about 20 µm, although some are only 5 µm in diameter and the red blood cells have to distort to go through. If we assume that the capillaries are evenly distributed at approximately 190/mm^2, then the total length of the capillaries in each cubic millimeter of muscle is about 190 mm. Since the volume occupied by 1 kg of muscle is about 10^6 mm^3, there are about 190 km (or over 100 miles) of capillaries in 1 kg of muscle! Taking the average diameter of a capillary to be 20 µm, the surface area of the capillaries in 1 kg of muscle is about 12 m^2—about the floor area of a typical bedroom!

Not all capillaries are carrying blood at any one time. In resting muscle only 2 to 5% of the capillaries are functional. The small arteries (arterioles) that supply the capillaries have circular cuffs of muscle (sphincters) that control the flow of blood in the capillary network (Fig. 8.4). When

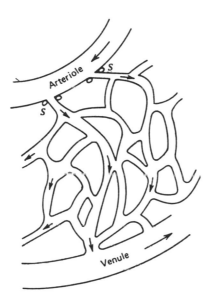

Figure 8.4. A small section of a capillary bed. A sphincter muscle (S) controls the blood flow into the capillaries.

there is a demand for blood, the cuffs relax allowing the muscle to get more blood and, hence, O_2.

Starling's law of capillarity describes the flow of fluids into and out of the capillaries Fluid movement through the capillary wall is the result of two pressures: the hydrostatic pressure, P, across the capillary wall and the osmotic pressure π. The capillary pressure varies from about 3.3 kPa (25 mm Hg) where the blood flows in at the arterial end to about 1.3 kPa (10 mm Hg) where the blood leaves the capillary at the venous end. The net osmotic pressure is estimated to be about $\pi = 3$ kPa (20 mm Hg) inside the capillary. Fluids flow out of the capillary at the arterial end and into the capillary at the venous end. If the capillary pressure should rise, for example due to trauma, more fluids would be forced into the tissues from the capillaries causing swelling, or edema, of the tissues. For more details see Chapter 6, *Osmosis and the Kidneys*.

8.3 Work Done by the Heart

In a typical adult each contraction of the heart muscles forces about 80 ml (about one-third of a cup) of blood through the lungs from the right ventricle and a similar volume to the systemic circulation from the left ventricle. In the process the heart does work. The volumes are not exactly equal for any one contraction, but over a period of time they pump the same amount.

The pressures in the two pumps of the heart are not the same (Fig. 8.5). In the pulmonary system the pressure is quite low because of the low resistance of the blood vessels in the lungs. The maximum pressure (systole), typically about 3 kPa (25 mm Hg), is about one-fifth of that in the systemic circulation. In order to circulate the blood through the much larger systemic network the left side of the heart must produce pressures that are typically about 16 kPa (120 mm Hg) at the peak (systole) of each cardiac cycle. During the resting phase (diastole) of the cardiac cycle the pressure is typically about 10.5 kPa (80 mm Hg). Note the greater thickness of muscles on the left side of the heart as seen in Fig.8.1a. The muscle driving the left ventricle is about three times thicker than that driving the right ventricle.

For reasons we discuss in Section 8.7, there is little loss of pressure until the blood reaches the arterioles and capillaries. Almost all of the pressure drop occurs across the arterioles and the capillary bed of the circulatory system (Fig. 8.5).

The power, or time rate at which energy is used, $\Delta E/\Delta t$, by a pump working at a constant pressure P is equal to the product of the pressure and the volume pumped per unit time, $\Delta V/\Delta t$.

$$\text{Power} = \Delta E/\Delta t = P\,\Delta V/\Delta t \qquad (8.1)$$

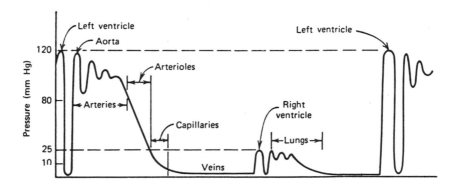

Figure 8.5. The pressure varies throughout the circulatory system. Note the low pressure in the veins and the relatively low pressure in the pulmonary system.

We can estimate the physical work done by the heart by multiplying its average pressure by the volume of blood that is pumped. Let us assume that the average pressure is about 13 kPa (100 mm Hg). If a volume of 8×10^{-5} m^3 (80 ml) of blood is pumped each second (a pulse rate of 60/min), the power is

$$(1.3 \times 10^4 \text{ Pa})(8 \times 10^{-5} \text{ m}^3/\text{s}) = 1.1 \text{ J/s or a power of } 1.1 \text{ W}.$$

Actually the pumping action takes place in less than one-third of the cardiac cycle and the heart muscle rests for over two-thirds of the cycle. Thus the power during the pumping phase is more than three times larger than the average value calculated above.

The heart, like all other engines, is not very efficient. In fact, it is typically less than 10% efficient, and the average power consumption of the heart is estimated to be over 10 W. Because of the lower blood pressure in the pulmonary system the power needed there is about one-fifth of that needed by the general circulation. During strenuous work or exercise the blood pressure may rise by 50% and the blood volume pumped per minute may increase by a factor of 5, leading to an increase of 7.5 times in the energy expended by the heart in each minute. We discuss the energy expenditure of enlarged hearts in Section 8.10.

8.7

PROBLEM

If the average power used by the heart is 10 W, what percentage of a 2500 kcal daily diet is used to operate the heart? (4.19 kJ = 1 kcal) [Answer: About 10%]

8.4 Blood Pressure and Its Measurement

One of the most common clinical measurements is of blood pressure. The first known experimental measurement of blood pressure was made in 1733 by the Rev. Stephen Hales in Great Britain. He bravely connected a 3 m (10 ft) vertical glass tube to an artery of a horse using the trachea of a goose as a flexible connection and a sharpened goose quill to puncture the artery! He found that the blood rose to an average height of 2.4 m (8 ft) above the heart.

During surgery and in intensive care wards, a direct measurement of blood pressure is frequently performed. Figure 8.6 shows a catheter placed in the arm; during catheterization of the heart, the catheter is advanced into the chambers of the heart. Every few minutes, the stop-cock is rotated so that a few milliliters of flushing solution pass through the catheter, thus preventing a clot from forming at the tip. The liquid enters the dome of the pressure transducer, where it pushes down on the metal diaphragm. The resulting bending of the diaphragm moves an armature, around which are wound fine strain-gauge wires. The wires are all under tension to begin with, but downward movement of the arma-ture increases the tension of two wires. The increased tension stretches these wires makes them narrower, and increases their resistance. The other two wires undergo slight compression. The decreased tension in the other two wires makes them fatter, and decreases their resistance. By placing the tension and compression wires in opposite arms of a bridge a voltage output is obtained that operates a meter or displays pulsatile

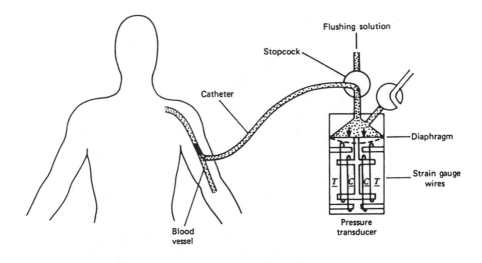

Figure 8.6. Direct blood pressure measurement. A hollow needle is inserted in the blood vessel and a catheter (hollow plastic tube) is threaded through the needle. The catheter transmits the blood pressure to the pressure transducer. The blood pressure deflects the diaphragm, causing a change of resistance to the four strain-gauge wires. The *T* wires undergo increased tension and the *C* wires undergo decreased tension.

waveforms on a video display or recorder. Any air bubbles in the system will result in errors in the shape of the recorded pressure waveform.

This direct method of measuring blood pressure is not necessary for routine purposes since reasonably accurate blood pressure measurements can be made by indirect means. The instrument that is commonly used is called a *sphygmomanometer*. It consists of a pressure cuff and gauge on the upper arm and a stethoscope placed over the brachial artery at the elbow (Fig. 8.7). The pressure cuff is inflated rapidly to a pressure sufficient to stop the flow of blood and then the air is released gradually. As the pressure in the cuff drops below the systolic blood pressure, the turbulent flow of blood squirting through the artery causes sound vibrations that can be heard in the stethoscope. They are called *Korotkoff* or *K sounds*. The pressure at which the K sounds are first heard indicates the systolic pressure level. As the pressure falls further, the K sounds become louder and then begin to fade. The point at which the K sounds die out or change indicates the diastolic pressure. The units on the pressure gauge are usually

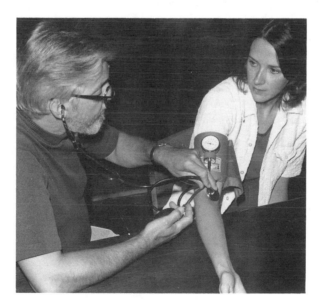

Figure 8.7. Blood pressure measurement using a sphygmomanometer. The arterial blood flow to the arm is blocked by an inflated cuff. As the air is gradually released, the stethoscope placed over the brachial artery is used to listen for the Korotkoff sounds. The pressures at which the sounds appear and change are noted on the gauge. (The "doctor" is one of the authors, JRC, and the "patient" is a former editorial assistant, Barbara Sandrik.)

mm Hg but they could easily be indicated in kPa [1 mm Hg = 0.133 kPa]. For individuals experienced in this technique the reproducibility (precision) of the systolic blood pressure measurement is usually within ±0.3 kPa (±2 mm Hg). The reproducibility of the diastolic measurement is not as good, ±0.7 kPa (±5 mm Hg). In addition, the accuracy is dependent on the obesity of the patient and other factors.

The pressure in the circulatory system varies throughout the body. Even in major arteries the pressure varies from one point to another because of gravitational forces. Figure 8.8a shows schematically direct measurements of blood pressure made on a standing person; open glass tube manometers are shown connected to arteries in the foot, upper arm, and head. In this situation the blood rises to essentially the same level in

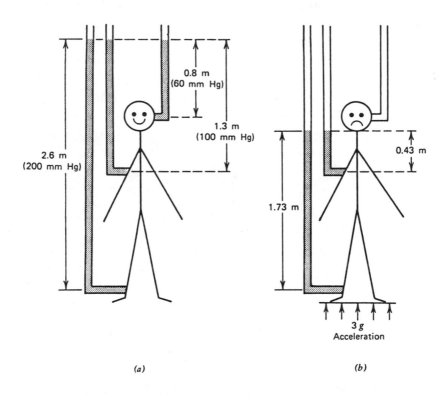

(a)   (b)

Figure 8.8. (a) If glass capillaries were connected to the arteries at different locations, the blood would rise to about the same level. If the body were horizontal, the blood pressure would be about the same at the three points instead of differing by a factor of over 3 as shown here. (b) If the body were accelerated upward at 3 g, the blood would not reach the brain and blackout would result.

all three manometers. The greater pressure P in the foot is due to the gravitational force (ρgh) produced by the column of blood (of height h) between the heart and foot added to the pressure at the heart (ρ is the density of the blood). Similarly, the decreased pressure in the head is due to the elevation of the head over the heart. Since mercury is about 13 times as dense as blood ($\rho_{Hg} = 13.6 \times 10^3$ kg/m^3, $\rho_{blood} = 1.04 \times 10^3$ kg/m^3) a column of mercury would be only one-thirteenth as high as a given column of blood. That is, if your blood pressure is 120/80 mm Hg (i.e., 120 mm Hg systolic and 80 mm Hg diastolic), it would be 1560/1040 measured in millimeters of blood. If the average pressure at your heart is 100 mm Hg, blood in a tube such as that shown in Fig. 8.8a would rise to an average height of 1300 mm or 1.3 m above your heart.

If gravity on earth suddenly became three times greater (i.e., $g \cong 30$ m/s^2), blood would rise only about 43 cm above the heart and it would not reach the brain of a standing person. This situation can be produced artificially by accelerating the body at a = 3 g in a head first direction (Fig. 8.8b). It can also occur in an airplane pulling out of a dive, causing the pilot to black out. These conditions also produce pooling of blood in the legs. Special tight-fitting suits that compress the legs have been designed to reduce this pooling.

A simple method for estimating the venous pressure at the heart is to observe the veins on the back of the hands. When the hands are lower than the heart the veins stand out because of increased venous pressure. As the hands are slowly raised above the level of the heart a point is reached at which the veins collapse; this indicates a pressure of 0 cm of blood. The height of the hand veins above the heart gives the venous pressure at the heart in centimeters of blood. Venous pressure normally averages 8 to 16 cm H$_2$O (or blood). A pressure in excess of 16 cm H$_2$O may indicate congestive heart failure (see Section 8.10).

8.8

PROBLEM

By how much does the blood pressure in your brain increase when you change from a standing position to standing on your head? Assume the same physical quantities as in Fig. 8.8a.
[Answer: The change is about 1.03×10^4 Pa, or 80 mm Hg]

8.5 Pressure across the Blood Vessel Wall (Transmural Pressure)

As indicated in Fig. 8.5, the greatest pressure drop in the cardiovascular system occurs in the region of the arterioles and capillaries. The capillaries have very thin walls (~1 μm) that permit easy diffusion of O_2 and CO_2. In order to understand why they do not burst we must discuss the law of Laplace, which tells us how the tension in the wall of a tube is related to the radius of the tube and the pressure inside the tube.

We can calculate the tension T in the wall of a long tube of radius R carrying blood at pressure P (Fig. 8.9a). The pressure is uniform on the wall, but we can mathematically divide the tube in half as shown in Fig. 8.9b. The force per unit length pushing upward is 2RP. There is a tension force T per unit length at each edge that holds the top half of the tube to the bottom half. Since the wall is in equilibrium, the force pushing the two halves apart is equal to the tension forces holding them together so

$$2T = 2RP \text{ or } T = RP. \tag{8.2}$$

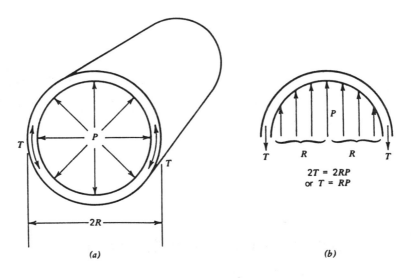

Figure 8.9. For a long tube of radius R with blood at pressure P (a), we can calculate the tension in the walls (b). The tension is very small for very small vessels, and thus their thin walls do not break.

Table 8.1. Typical Pressures and Tension in Blood Vessels

	Mean Pressure		Radius (cm)	Tension (N/m)
	(mm Hg)	(kPa)		
Aorta	100	13	1.2	156.0
Typical artery	90	12	0.5	60.0
Small capillary	30	4	6×10^{-4}	0.024
Small vein	15	2	2×10^{-2}	0.40

For a very small radius (e.g., in a capillary) the tension is correspondingly very small.

Table 8.1 gives some typical pressures and tensions in the blood vessels. For example, the tension in the wall of the aorta is about 156 N/m, while the tension in a capillary wall is only about 24×10^{-3} N/m. For comparison, a single layer of toilet tissue can withstand a tension of about 50 N/m. This tension is about 3000 times greater than a tension which would rupture the capillary.

8.9

PROBLEM

Show with a calculation and explain why arteries with small diameters can have thinner walls than arteries with large diameters carrying blood at the same pressure.

8.6 Bernoulli's Principle Applied to the Cardiovascular System

You are probably familiar with the Bernoulli principle even though you might not give Bernoulli credit for it. Whenever there is a rapid flow of a fluid such as air or water, the pressure is reduced at the edge of the rapidly moving fluid. For example, the rapid flow of the water in the shower causes a reduced pressure in the vicinity of the shower curtain and it pulls in toward the water. Similarly, when the window in a moving car is first rolled down the reduced pressure caused by the air moving rapidly outside the window causes objects to fly out of the window.

Bernoulli's principle is based on the law of conservation of energy. Pressure in a fluid is a form of potential energy, PE, since it has the ability to perform useful work. In a moving fluid there is kinetic energy, KE, as a result of the motion. This kinetic energy can be expressed in terms of energy per unit volume, such as joules per cubic meter. Since 1 joule (J) = 1 Nm, then 1 J/m^3 = 1 (Nm)/m^3 or 1 N/m^2 = 1 Pa, the unit for pressure in the SI system. If fluid is flowing through the frictionless tube shown in Fig. 8.10, the velocity increases in the narrow section and the increased kinetic energy of the fluid is obtained by a reduction of the potential energy of the pressure in the tube. As the velocity reduces again on the far side of the restriction, the kinetic energy is converted back into potential energy and the pressure increases again, as indicated on the manometers.

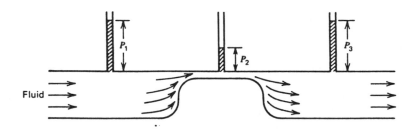

Figure 8.10. As the velocity of the fluid increases in the narrow section of the tube, part of the potential energy (pressure) is converted into kinetic energy so there is a lower pressure, P$_2$, in this section. P$_2$ is less than P$_1$ and P$_3$.

We can calculate the average kinetic energy per unit volume of 10^{-3} kg (10^{-6} m^3 = 1 ml) of blood as it leaves the heart. Since the average velocity is about 0.3 m/s, the kinetic energy of this mass of blood is:

$$KE = (1/2) \ mv^2 = (1/2) \times (10^{-3}) \times (0.3)^2 = 4.5 \times 10^{-5} \text{ J.} \quad (8.3)$$

Since the volume involved is 10^{-6} m^3, this can be thought of as an energy per unit volume of 45 J/m^3 of blood leaving the heart. From the reasoning in the previous paragraph, this energy density is equivalent to a pressure of 45 Pa or about 0.4 mm Hg. However, during heavy exercise the velocity of the blood being pumped by the heart may be five times its average value during rest, and during the peak of the heartbeat the kinetic energy factor can have a pressure equivalent of 10 kPa (75 mm Hg) and can represent 30% of the total work of the heart.

8.7 How Fast Does Your Blood Flow?

As the blood moves away from the heart, the arteries branch and re-branch many times to carry blood to the various tissues. The smallest blood vessels are the capillaries. As discussed in Section 8.2, they are very small (~20 μm in diameter) and there are millions of them. There are so many carrying blood that their total cross-sectional area is equivalent to that of a tube almost 0.3 m in diameter! Total cross-sectional areas of the vessels in the circulatory system are shown schematically in Fig. 8.11.

As the blood goes from the aorta into the smaller arteries and arterioles with greater total cross-sectional areas, the velocity of the blood

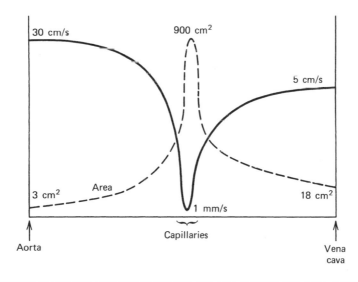

Figure 8.11. The dashed curve shows schematically the change in the cross-sectional area of the circulatory system. The velocity of blood flow (solid line) decreases as the total cross-sectional area increases. The total cross-sectional area is obtained by adding the areas of all blood vessels at a given distance from the heart. Note that the vena cava returning the blood to the heart has a much larger cross-sectional area than the aorta.

decreases much as the velocity of a river decreases at a wide portion. Figure 8.11 also shows schematically the velocity of blood flow in the different portions of the circulatory system. Notice that the blood velocity is related in an inverse way to the total cross-sectional area of the vessels carrying the blood. The average velocity in the aorta is about 0.3 m/s; that in a capillary is only about 10^{-3} m/s (1 mm/s). It is in the capillaries that the exchange of O_2 and CO_2 takes place, and this low velocity allows time for diffusion of the gases to occur.

You are undoubtedly aware of the characteristic of a liquid called viscosity (η). The syrup you pour on your pancakes pours at a rate different from that of the cream you put in your coffee or the water you pour into a glass. The slipperiness or ease with which a fluid pours is an indication of its viscosity. The SI unit for viscosity is the pascal-second (Pa·s). The viscosity of water is about 10^{-3} Pa·s at 20 C. The viscosity of thick syrup may be 100 Pa·s. The viscosity of blood is typically 3 to 4×10^{-3} Pa·s, but depends on the percentage of erythrocytes in the blood (known as the hematocrit). As the hematocrit increases, the viscosity increases (Fig. 8.12). Persons with the disease *polycythemia vera* in which there is an over-production of erythrocytes have a high hematocrit and often have circulatory problems. The viscosity of the blood also depends on temperature. As blood gets colder, the viscosity increases and this further reduces the blood supply to cold hands and

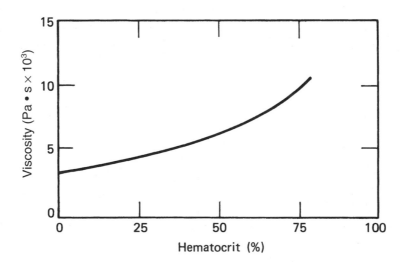

Figure 8.12. As the percent of red blood cells in the blood increases (higher hematocrit), the viscosity increases, decreasing the flow rate.

feet. A change from 37 C to 0 C increases the viscosity of blood by a factor of 2.5. Cigarette smokers generally have a higher hematocrit than non-smokers. This is likely the result of the fact that smokers breathe in 250 ml of carbon monoxide (CO) from each pack of cigarettes. The CO reduces the ability of the erythrocytes to carry O_2 and the body compensates by producing more erythrocytes. The greater the hematocrit, the greater is the viscosity, which may lead to more cardiovascular disease such as stroke and heart attacks.

In addition to viscosity, other factors affect the flow of blood in the vessels: the pressure difference from one end to the other, the length of the vessel, and its radius. In order to understand the laws that control the flow of blood in the circulatory system, Poiseuille in the nineteenth century studied the flow of water in tubes of different sizes. The results of his experiments are summarized in Fig. 8.13. Poiseuille's law states that the flow through a given tube depends on the pressure difference from one end to the other ($P_A - P_B$), the length L of the tube, the radius R of the tube, and the viscosity η of the fluid. If the pressure difference is doubled, the flow rate also doubles. The flow varies inversely with the

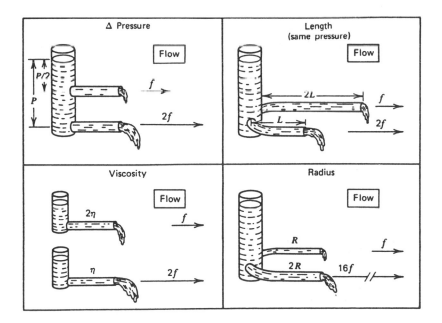

Figure 8.13. Poiseuille's findings. The flow rate through a tube depends on the pressure difference from one end of the tube to the other, the length of the tube, the viscosity of the fluid, and the radius. The radius has the largest influence on flow rate.

length and with the viscosity; if either is doubled, the flow rate is reduced by one-half. Poiseuille's most surprising discovery was the way that the flow rate depends on the radius of the tube. As he expected, the flow rate increased as the radius of the tube increased; what was surprising was how rapidly the flow rate increased with small increases in the radius. For example, if the radius is doubled, the flow rate increases by 2^4 or a factor of 16. When all of these variables are put together with a constant to keep the units working correctly we get Poiseuille's equation:

$$\text{Flow rate} = (P_A - P_B)(\pi/8)(1/\eta)(R^4/L) \qquad (8.4)$$

In SI units, the flow rate will be in m^3/s if $P_A - P_B$ is in N/m^2, η is in Pa·s, and R and L are in m.

Poiseuille's law applies to rigid tubes of constant radius. Since the major arteries have elastic walls and expand slightly at each heartbeat, blood flow in the circulatory system does not obey the law exactly. In addition, the blood's viscosity changes slightly with flow rate; however, this effect is negligible.

Even though the total cross-sectional area of the arterioles is many times greater than that of the aorta, most of the pressure drop occurs across the arterioles because of the great flow resistance produced by the R^4 factor. The next largest drop is across the capillaries (Fig. 8.5).

8.10

PROBLEM

What is the approximate velocity of blood in the capillaries?

8.11

PROBLEM

If the radius of an arteriole changed from 50 to 40 μm, how much would the flow rate through it change? Does the flow rate increase or decrease? [Answer: Decrease by about 60%]

8.8 Blood Flow—Laminar and Turbulent

You have probably seen both a slow, smooth, quietly flowing river and a rapid, turbulent, noisy river. The first type of river is analogous to the laminar or streamline flow that is present in most blood vessels. The second is similar to the turbulent flow found at a few places in the circulatory system; for example, where the blood is flowing rapidly past the heart valves.

An important characteristic of laminar flow is that it is silent. If all blood flow were laminar, information could not be obtained from the heart with a stethoscope. The heart sounds heard with a stethoscope are caused mostly by turbulent flow. During a blood pressure measurement, the constriction produced by the pressure cuff on the arm produces turbulent flow and the resulting vibrations can be detected with a stethoscope on the brachial artery.

In laminar flow the blood that is in contact with the walls of the blood vessel is essentially stationary, the layer of blood next to the outside layer is moving slowly, and successive layers move more rapidly just as the water in the middle of a quiet stream moves more rapidly than the water along the banks (Fig. 8.14a). This behavior has an effect on the distribution of red blood cells in the circulatory system.

The red blood cells in an artery are not distributed uniformly; there are more in the center than at the edges (Fig. 8.14b). This produces two effects. When blood enters a small vessel from the side of a main vessel, the percentage of red blood cells in that blood (the hematocrit) will be slightly less than in the blood in the main vessel because of the "skimming" effect. The second effect is more important. Because the plasma along the vessel walls is moving more slowly than the red blood cells, the blood in the extremities has a greater percentage of red blood cells than when it left the heart. This causes an increase in the hematocrit in the hands and feet of approximately 10% over the hematocrit of the whole blood.

If you gradually increase the velocity of a fluid flowing in a tube by reducing the radius of the tube, it will reach a critical velocity v_c when laminar flow changes into turbulent flow (Fig. 8.15). The critical velocity will be lower if there are restrictions or obstructions in the tube. Osborne Reynolds studied this property in 1883 and determined that the critical velocity is proportional to the viscosity η of the fluid and is inversely proportional to the density ρ of the fluid and the radius R of the tube, that is, $v_c = K\eta/\rho R$. The constant of proportionality K is called the Reynold's number, and it is approximately equal to 1000 for many fluids, including blood, flowing in long straight tubes of constant diameter.

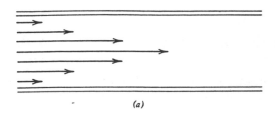

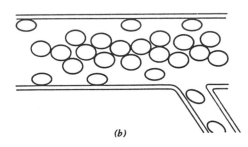

Figure 8.14. Blood flow in the vessels. (a) In the laminar flow in most of the vessels there is a greater velocity at the center as indicated by the longer arrow. (b) The distribution of red blood cells is not uniform; they are more dense at the center so the blood that flows into small arteries has a smaller percentage of red blood cells than the blood in the main artery.

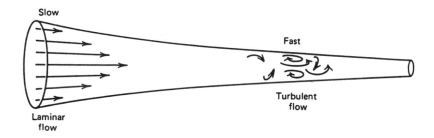

Figure 8.15. If the fluid is flowing in a long tapering tube, the velocity will gradually increase to the point where it exceeds the critical velocity v_c, producing turbulent flow.

If there are bends or obstructions, the Reynold's number becomes much smaller. In the aorta, which has a radius of about 0.01 m (1 cm) in adults, the critical velocity is

$$v_c = \frac{(1000)\,(4 \times 10^{-3}\ \text{Pa} \cdot \text{s})}{(10^3\ \text{kg/m}^3)(10^{-2}\ \text{m})} = 0.4\ \text{m/s}.$$

The velocity in the aorta ranges from 0 to 0.5 m/s, and thus the flow is turbulent during part of the systole. During heavy exercise the amount of blood pumped by the heart may increase four or five times and the critical velocity will be exceeded for a longer period of time. The heart sounds of a person doing heavy exercise are different from those of a person at rest.

Laminar flow is more efficient than turbulent flow. This is illustrated graphically in Fig.8.16a. The slope of the curve in the laminar flow region is greater than that in the turbulent flow region. That is, a given

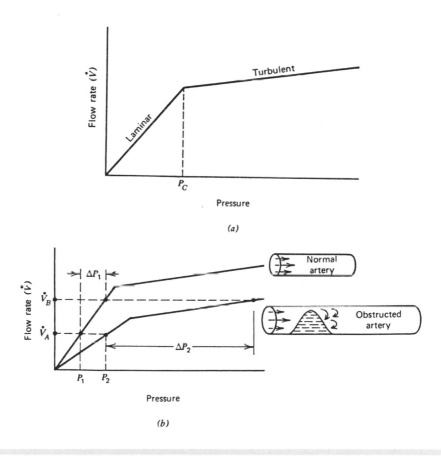

Figure 8.16. (a) When the flow in a tube becomes turbulent (at pressure P_c), the slope of the flow rate versus pressure decreases so that compared to laminar flow a greater increase in pressure is necessary to obtain a given increase in flow rate. (b) In an obstructed artery the pressure P_2 needed to produce a given flow rate $\dot{V}_A$ is greater than the pressure P_1 to produce the same flow rate in a normal artery. In addition, if the heart is called upon to increase the flow rate from $\dot{V}_A$ to $\dot{V}_B$, the turbulence produced in the obstructed artery requires a much larger pressure increase (ΔP_2 vs. ΔP_1) and thus greater effort from the heart. (Adapted from I. W. Richardson and E. B. Neergaard, *Physics for Medicine and Biology*, Wiley-Interscience, New York, 1972, pp. 46–47.)

increase in pressure causes a greater increase in the laminar flow rate than in the turbulent flow rate. The reduction in efficiency is apparent in the blood flow through an artery with an obstruction (Fig. 8.16b). For the flow rate $\dot{V}_A$, a pressure of P_1 is needed for the normal artery, and a somewhat higher pressure P_2 is needed for the obstructed artery. If both arteries are required to deliver an increased flow rate $\dot{V}_B$, the normal artery requires a pressure increase of ΔP_1. The obstructed artery, because of turbulent flow, requires a much greater pressure increase of ΔP_2 and a proportionately greater amount of work.

PROBLEM 8.12

An artery with a 3 mm radius is partially blocked with plaque; in the constricted region the effective radius is 2 mm and the average blood velocity is 0.5 m/s.
(a) What is the average velocity of the blood in the unconstricted region? [Answer: 0.22 m/s]
(b) Would there be turbulent flow in either region? [Answer: No]
(c) For the blood in the constricted region, find the equivalent pressure due to the kinetic energy of the blood. [Answer: 130 Pa, or 1 mm Hg]

8.9 Heart Sounds

An experienced cardiologist with good hearing can obtain much diagnostic information from the heart sounds. The heart sounds heard with a stethoscope are caused by vibrations originating in the heart and the major vessels. The opening and closing of the heart valves contribute greatly to the heart sounds; turbulent flow occurs at these times and some of the vibrations produced are in the audible range. Figure 8.17 shows the sounds heard with a stethoscope from a normal heart. Other sounds may be heard if the heart is not normal. Murmurs may be produced if there is a constriction that causes turbulent flow during part of the cardiac cycle. For example, if the aortic valve is narrow (aortic valve stenosis) blood flow through it during systole will cause a murmur.

The amount and quality of the sound heard depend on the design of the stethoscope as well as on its pressure on the chest, its location, the orientation of the body and the phase of the breathing cycle. There are optimum positions for hearing the various heart sounds with a stethoscope.

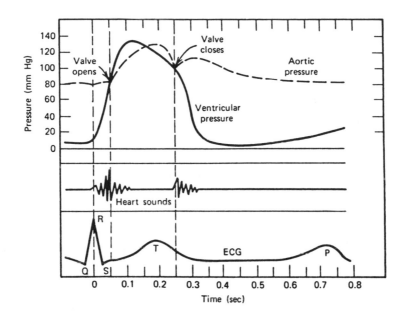

Figure 8.17. The time relationships of the electrocardiogram (ECG) (bottom), heart sounds (phonocardiogram), and left ventricular and aortic blood pressure (top). The ventricle begins to contract at time 0, and the aortic valve opens when the ventricular pressure just exceeds the aortic pressure. This contributes to the first heart sound. The closing of the aortic valve contributes to the second heart sound. The first sound is normally longer and louder than the second sound. (Adapted from Sher, A. M. in T. C. Ruch and H. E. Patton (Eds.), *Physiology and Biophysics*, 19th ed., © W. B. Saunders Company, Philadelphia, 1965, p 557.)

In general, sound is not transmitted well from a liquid to air, and thus heart sounds are not heard well if the sound must travel through the lung.

The sounds from a normal heart are in the frequency range of 20 to about 200 Hz. This is not the most sensitive range of the human ear (see Chapter 11, *Physics of the Ear and Hearing*). The sensitivity of the ear is very poor at low frequencies; to be heard, a sound at 20 Hz must be about 10,000 times more intense than a sound at 200 Hz. A normal heart produces some sounds that cannot be heard with a good stethoscope, even under optimum conditions. However, it is possible to electronically amplify these heart sounds to listen to them directly or to record them. Phonocardiography is the graphic recording of heart sounds (Fig. 8.17). The electronic amplifiers used in phonocardiography have a much different response than the human ear so the recordings do not correspond well with what the cardiologist hears. Similarly, an electronically amplified stethoscope distorts the sounds that the physician is accustomed to hearing.

8.10 The Physics of Some Cardiovascular Diseases

Heart disease is the number one cause of death in the United States. Because of the many physical aspects of the cardiovascular system, heart diseases often have a physical component. Many of these diseases, for example, increase the work load of the heart or reduce its ability to work at a normal rate.

The work done by the heart is roughly the tension of the heart muscle times how long it acts. Anything that increases the muscle tension or how long it acts will increase the work load of the heart. For example, high blood pressure (hypertension) causes the muscle tension to increase in proportion to the pressure. A fast heart rate (tachycardia) increases the work load since the heart pumps more blood per minute.

The heart disease that causes the most deaths is heart attack. A heart attack is caused by a blockage of one or more coronary arteries to the heart muscle. That portion of the heart muscle without a blood supply dies (an infarct). The blockage does not always immediately affect the electrical signals that control the heart's beating action, and thus a person with a recent heart attack may still have a normal electrocardiogram (ECG).

During and after a heart attack the ability of the heart muscle to pump blood to the body is seriously impaired. To reduce the work load of the heart, bed rest and O_2 therapy are prescribed. Giving O_2 increases the O_2 in the blood so that less blood needs to be pumped to the tissues. This O_2 is probably most beneficial to the heart muscle itself. There are often alternate arteries (anastomoses) for blood to get to muscles. These anastomoses can provide some O_2 to the blocked portion. One of the purposes of a regular exercise program is to stress the cardiovascular system enough to keep these anastomoses open.

Another common heart disease is congestive heart failure. The cause of this disease is not as well understood as the cause of heart attack. It is characterized by an enlargement of the heart and a reduction in the ability of the heart to provide adequate circulation.

For an enlarged heart we can apply the law of Laplace. If the radius of the heart is doubled, the tension of the heart muscle must also be doubled if the same blood pressure is to be maintained. However, since the heart muscle is stretched, it may not be able to produce sufficient force to maintain normal circulation. The stretched heart muscle is also much less efficient than normal heart muscle; that is, it consumes much more O_2 for the same amount of work.

The medical treatment for congestive heart failure is to reduce the work load of the heart. A dramatic approach is to replace the heart surgically. Many heart transplants have been successful. The number of heart transplants in 1998 was about 2300. At the end of a year nearly 90% were still alive. Typically, at any one time, about 3700 patients are waiting for a suitable match for a cadaver heart.

Patients with a condition in which the heart's electrical signals are inadequate to stimulate heart action have been greatly helped by modern technology. They have received artificial pacemakers to regulate the heartbeat.

Another device that has helped heart patients is the artificial heart valve. Heart valve defects are of two types: the valve either does not open wide enough (*stenosis*) or it does not close well enough (*insufficiency*). In stenosis the work of the heart is increased because a large amount of work is done against the obstruction of the narrow opening, and the blood supply to the general circulation is reduced. In insufficiency some of the pumped blood flows back into the heart so that the volume of circulated blood is reduced. Both types of defective valves can be replaced by artificial valves. Several designs are available (Fig. 8.18). Compatibility between artificial valves and the blood is still a problem. These valves sometimes cause clotting.

Cadaver heart valves are also used as replacements for defective valves. A heart valve is primarily cartilage tissue with relatively few living cells. Before it is transplanted, a cadaver valve is sterilized with radiation from an electron accelerator that is used for treating cancer. It is now common to use valves from pig hearts.

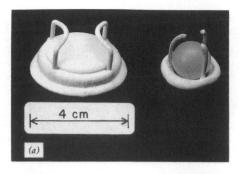

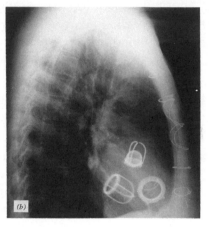

Figure 8.18. (a) Two of the several artificial heart valves used routinely. Such a valve is sutured into the heart to allow blood to flow upward only. (b) An x-ray of a patient showing three artificial heart valves in place. At the time the x-ray was taken in 1976, the valves had been in place 6.5 years and the patient was well. (Courtesy of Dr. William Young, University of Wisconsin, Madison.)

Many cardiovascular diseases involve the blood vessels. We now discuss the physics of a few of these diseases.

An aneurysm is a weakening in the wall of an artery resulting in an increase in its diameter (Fig. 8.19). The increased diameter increases the tension in the wall proportionately. If it were not for the supporting action of the surrounding tissue, the wall would blow out the way a bicycle inner tube does under similar conditions. If an aneurysm does rupture, it is often fatal—especially if the rupture is in the brain, a type of cerebrovascular accident (CVA).

A more common vessel problem is the formation of sclerotic plaques on the walls of an artery. The plaques can cause turbulent flow and produce a noticeable murmur. The narrowing of the artery will cause

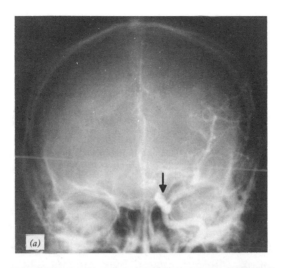

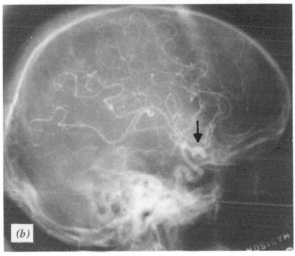

Figure 8.19. X-rays of the skull from the front (a) and the side (b) showing an aneurysm (arrow). A dye that absorbs x-rays has been injected into the arteries to make them visible.

an increase in the blood velocity in that region with a decrease in wall pressure because of the Bernoulli effect. The plaque may dislodge and travel with the blood until it lodges in a smaller artery. This blockage will shut off the blood supply to the affected part; if it is in the brain, it will produce a stroke, another type of cerebrovascular accident.

A disease that is clinically not as serious as aneurysms and plaques but that often causes embarrassment is varicose veins. Varicose veins can be more than a cosmetic problem since they can develop complications.

These enlarged surface veins in the leg result from a failure of the one-way valves in the veins. Consider the blood flow in the lower legs and feet of an erect person. The pressure in a leg vein is approximately 12 kPa (90 mm Hg, or 115 cm of blood) as the result of the column of blood above it. During walking or other leg exercise, the contraction of the muscles forces the venous blood toward the heart. This action of the muscles on the blood is called the venous pump or muscle pump. At various points along the veins there are one-way flaps or valves that prevent the blood from going back. The action of the muscle pump and the valves results in a venous pressure of about 3 kPa (20 mm Hg) during exercise. If these valves become defective and let the blood run back down, the blood will pool in the vein and the vein will become varicose. Varicose veins may be aggravated by conditions that restrict the return of the blood to the heart. The additional abdominal weight during pregnancy may restrict venous return. However, some experiments have shown that a pressure cuff above the knee inflated to 12 kPa (90 mm Hg) is no hindrance to the action of the muscle pump on the internal veins. The standard treatment for varicose veins is surgical removal of the offending vessels. There are usually adequate parallel veins to carry the blood back to the heart.

8.13

PROBLEM

Explain how the venous blood returns from the feet to the heart of a standing person.

8.11 Some Other Functions of Blood

Although we have emphasized the role of blood in gas exchange, an equally important function of the blood is to carry the body's liquid wastes to the kidneys. The details of kidney function were discussed in Chapter 6, *Osmosis and the Kidneys*. By filtration of the blood, the kidneys keep the makeup of the blood very constant despite large fluctuations in our diet. The kidneys are well vascularized in order to filter the blood. Normally 1 to 1.5 liters of blood (one-fifth to one-fourth of the cardiac output) flows through the kidneys each minute. This is far in excess of the amount needed to supply nutrients and oxygen to the kidneys. If a severe blood loss occurs, the kidney flow may drop to 0.25 liter/min to permit the blood to be used elsewhere. Although artificial kidneys (dialysis units) are used by thousands of people they fall far short of the kidneys in their ability to regulate body components. They are often lifesaving while the patient waits for a donor kidney

The blood plays an important role in distributing and dissipating heat in the body (see Chapter 2, *Energy, Heat, Work, and Power of the Body*). The venous blood returning from the limbs can be routed close to the skin to increase heat losses in warm weather. In cold weather it can be routed internally close to the artery carrying blood to the limb; the cool venous blood takes up some of the heat from the warm arterial blood and carries it back to the heart. This counter-current principle keeps down heat losses from the extremities and from the skin in cold weather.

Blood is also involved in the male erection. On stimulation, arterial blood flows into the penis causing it to enlarge and become rigid due to the fluid pressure. The venous return is slow— the erection helps to block the return of blood to the veins. The blood pressure in the penis during an erection is about the same as the systolic pressure. The same principle keeps plants erect. Their rigidity is due to internal fluid pressure—if you do not water them, the fluid pressure drops and they wilt.

9

Electrical Signals from the Body

This chapter is primarily about *bioelectricity* and *biomagnetism*—electrical and magnetic signals generated by the body. Bioelectricity, such as the electrocardiogram, was known nearly a century before biomagnetism was discovered, such as the magnetocardiogram, magnetic signals generated by the heart. Much more is known about bioelectricity. Most of this chapter is about bioelectricity. Biomagnetism, measuring weak magnetic fields from the body, is described in Section 9.7. *Electrobiology* and *magnetobiology* refer to applications of external electricity and magnetism to the body, which are not discussed in this chapter.

People were aware of electric fish (torpedoes and eels) many centuries before electricity was scientifically studied. Luigi Galvani, an Italian anatomist, observed the first evidence that electricity played a role in muscle action, in 1786. He found that if two pieces of different metals were connected and their open ends were touched to different parts of the muscle of a dead frog, the frog's muscles would contract. He thought that the dead frog generated the electric stimulus. In fact, the muscle was stimulated by a weak electric current produced accidentally by a crude battery consisting of two metals as electrodes and body fluid as the electrolyte. Alessandro Volta investigated the phenomena and, in the process, invented the battery—one of the most important inventions in the history of physics. It was the first source of a steady electric current.

History does not record who was the first to propose that electrical pulses control muscles as well as provide signals to and from the brain. Sensitive instruments (galvanometers) to measure the weak electrical potentials from the heart were not available until about a century after Galvani's discovery. Jacques D'Arsonval invented the galvanometer, a sensitive instrument for measuring current, in 1880. Its slow response could not record the shape of the brief electrical signal when the heart contracted. The detection of the very short electrical signal from a neuron, an action potential, which lasts about a few milliseconds, had to await the invention of the oscilloscope in the 20th century.

The electricity generated inside the body serves for the control and operation of nerves, muscles, and organs. Essentially all functions and activities of the body involve electricity in some way. The forces of muscles are caused by the attraction of opposite electrical charges. The action of the brain is basically electrical. All nerve signals to and from the brain involve the flow of electrical currents.

The nervous system plays a fundamental role in nearly every body function. The brain, basically a central computer, receives internal and external signals and (usually) makes an appropriate response. The information is transmitted as electrical signals along various nerves. This efficient communication system can handle many millions of pieces of information at one time.

In carrying out the special functions of the body, many electrical signals are generated. These signals are the result of the electrochemical action of certain types of cells. By selectively measuring the appropriate signals we can obtain useful clinical information about particular body functions. In this chapter we discuss some of these electrical signals. Recorded electrical signals from the heart, the electrocardiogram (ECG); from the brain, the electroencephalogram (EEG); and from muscles, the electromyogram (EMG) are the best known. We also discuss less familiar electrical signals, such as from the retina, the electroretinogram (ERG) and from the eye muscles, the electrooculogram (EOG). Section 9.7 discusses magnetic signals from the heart, the magnetocardiogram (MCG) and from the brain, the magnetoencephalogram (MEG).

Various medical specialists are involved in the diagnosis and treatment of malfunctions of the body's internal electrical system. If the problem involves any part of the nervous system, it is diagnosed and treated by a *neurologist*, an M.D. who has had three or more years of special training in the study of the nervous system. If the problem requires surgery, it is usually handled by a *neurosurgeon*, an M.D. who specializes in surgery of the nervous system and has had three or more years of training in this area of surgery. Since much of neurosurgery involves the brain, these specialists are sometimes called brain surgeons. *Neuroradiologists* are M.D.'s who have taken a three-year residency in diagnostic radiology followed by another year in neuroradiological specialization. Pediatric neurology, a subspecialty of pediatrics, deals with nerve problems in infants and children. Electromyogram tests are usually performed and interpreted by *physiatrists*, M.D.'s who have taken residencies in physical medicine. *Cardiologists*, M.D. specialists in the study and treatment of heart disease, deal with diseases of the heart. *Psychiatrists* are M.D.'s who specialize in the diagnosis, prevention, and treatment of emotional illness and neural disorders. *Clinical psychologists* are Ph.D.'s who have studied behavior and also specialize in the diagnosis and treatment of mental illness; however, they cannot treat with drugs; they may use hypnosis.

9.1
PROBLEM

List five electrical signals from the body that are sometimes recorded.

9.1 The Nervous System and the Neuron

The nervous system can be divided into two parts—the central nervous system and the autonomic nervous system. The central nervous system consists of the brain, the spinal cord, and the peripheral nerves. Nerve fibers (*neurons*) that transmit sensory information to the brain or spinal cord are called *afferent nerves*. Nerve fibers that transmit information from the brain or spinal cord to the appropriate muscles and glands are called *efferent nerves*. The autonomic nervous system controls various internal organs such as the heart, intestines, and glands. The control of the autonomic nervous system is essentially involuntary.

The brain is exceedingly complicated and not well understood. It is the body's most important organ and is given special protection. It is surrounded by three membranes within the protective skull and because it "floats" in the shock-absorbing cerebrospinal fluid (CSF), the 1.5 kg brain has the effective weight of a 50 g mass. The brain is connected to the spinal cord, which is also surrounded by CSF and protected by the bone of the spinal column.

The basic structural unit of the nervous system is the *neuron* (Fig. 9.1), a nerve cell specialized for the reception, interpretation, and transmission of electrical messages. There are many types of neurons. Basically, a neuron consists of a cell body that receives electrical mes-

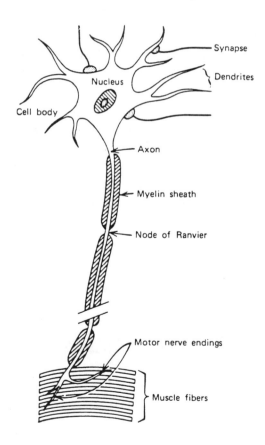

Figure 9.1. Schematic of a motor neuron.

sages from other neurons through contacts called *synapses* located on the dendrites or on the cell body. The *dendrites* are the parts of the neuron specialized for receiving information from stimuli or from other cells. If the stimulus is strong enough, the neuron transmits an electrical signal outward along a fiber called an *axon*. The axon, or nerve fiber, which may be as long as 1 m, carries the electrical signal to muscles, glands, or other neurons.

9.2 Electrical Potentials of Nerves

The ability of neurons to receive and transmit electrical signals is fairly well understood. In this section we discuss the electrical behavior of neurons. Much of the early research on the electrical behavior of nerves was done on the giant nerve fibers of squid. The conveniently large diameter (~1 mm) of these nerve fibers allows electrodes to be readily inserted or attached for measurements.

Across the surface (or membrane) of every neuron is an electrical potential (voltage) difference due to a net negative charge on the inner surface of the membrane and a net positive charge on the outer surface. The net charge is a result of a complicated interplay of negative and positive ions. The neuron is said to be *polarized*. The inside of the cell is typically 60 to 90 mV more negative than the outside. This potential difference is called the *resting potential* of the neuron. Figure 9.2 shows schematically the typical concentration of various ions inside and outside the membrane of an axon. When the neuron is stimulated, a large momentary change in the resting potential occurs at the point of stimulation. This potential change, called the *action potential*, propagates along the axon. The propagation of an action potential is the major method of transmission of signals within the body. The stimulation may be caused by various physical and chemical stimuli such as heat, cold, light, sound, and odors. If the stimulation is electrical, only about a 20 mV change across the membrane is needed to initiate the action potential.

We give a simple explanation for the resting potential by using a model in which a very thin membrane separates an initial concentrated neutral solution of KCl (left side of Fig. 9.3a) from a less concentrated solution (right side of Fig. 9.3a). The KCl in solution forms K^+ ions and Cl^- ions, shown in Fig. 9.3 by (+) and (−), respectively. We assume that the membrane is permeable to chloride ions and impermeable to

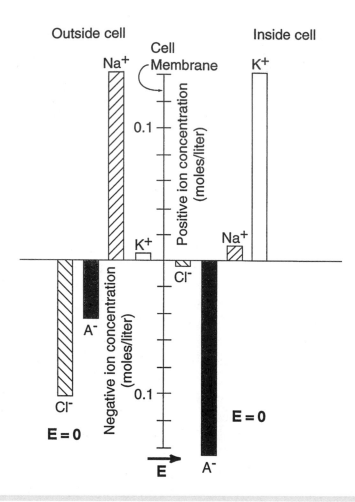

Figure 9.2. Typical concentrations in moles per liter of K^+, Na^+, Cl^-, and large protein ions (A^-) inside and outside a cell. The inside of the cell is more negative than the outside by about 60 to 90 mV. Note the electric field shown as **E** exists only at the membrane while the regions away from the membrane are neutral in charge.

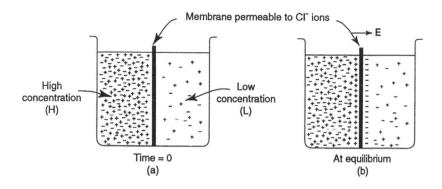

Figure 9.3. A model used to explain the existence of the resting potential.

potassium ions. From the initial starting conditions, the chloride ions will diffuse through the membrane giving a net motion of negative charge from left to right. Since the potassium ions cannot move through the membrane, there will quickly develop a deficit of chloride ions on the left of the membrane, leaving a positive charge there. The chloride ions that have moved through the membrane will provide a negative charge layer on the right (Fig. 9.3b). The separated charges behave like a simple charged capacitor, giving an electric field E pointing from left to right. As this field increases, it eventually completely retards the flow of the chloride ions. Accompanying the electric field is a potential difference across the membrane providing a simple explanation of the resting potential. The charges are located very close to the membrane, of the order of a few nm, the extent depending upon the ion concentration. The bulk solution beyond the ions at the membrane is conducting and the ions can move, thereby ensuring the electric field is zero in these regions, resulting in no net charge in any bulk volume. This is evident in Fig. 9.2 where the ion concentrations total zero both inside and outside of the regions away from the membrane.

While the discussion above provides a simple model for the existence of a potential difference across a membrane, the resting potential of a nerve is obviously more complicated. Fig. 9.2 shows the main ion concentrations with the A^- (protein) ions being unable to pass through the membrane. Measurements show that the chloride ions act like the equilibrium case discussed above while the potassium ions slowly leak out of the cell while the sodium ions leak in. An active chemical process "pumps" the potassium ions back into the cell and the sodium ions back out of the cell, providing the equilibrium concentration of Fig. 9.2. The potassium and sodium ions play major roles in the stimulation and propagation of the action potential.

Figure 9.4 shows schematically how the action potential propagates along an axon. Graphs of the potential measured between point P and the outside of the axon are also shown. This axon has a resting potential of about −80 mV (Fig. 9.4a). If the left end of the axon is stimulated, the membrane walls become porous to Na$^+$ ions and these ions pass through the membrane, causing it to depolarize. The inside momentarily goes positive to about 50 mV. The reversed potential in the stimulated region causes ion movement, as shown by the arrows in Figure 9.4b, which in turn depolarizes the region to the right (Fig. 9.4c, d, and e). Meanwhile the point of original stimulation has recovered (re-polarized) because K$^+$ ions have moved out to restore the resting potential (Fig. 9.4c, d, and e). The voltage pulse is the action potential. For most neurons and muscle cells, the action potential lasts a few milliseconds; however, the action potential for cardiac muscle may last from 150 to 300 ms (Fig. 9.5). An axon can transmit an action

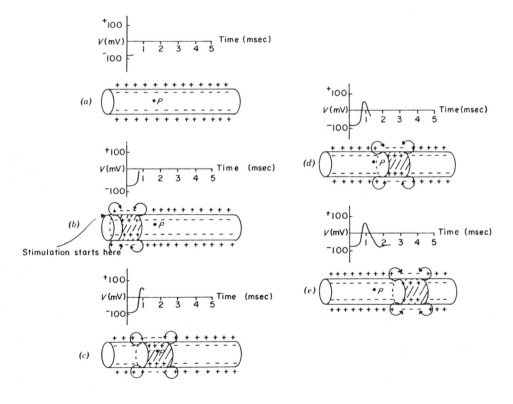

Figure 9.4. The transmission of a nerve impulse along an axon. The voltage pulse moving along the nerve is the action potential.

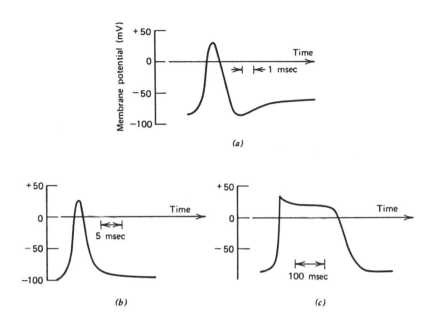

Figure 9.5. Waveforms of the action potentials from (a) a nerve axon, (b) a skeletal muscle cell, and (c) a cardiac muscle cell. Note the different time scales.

potential in either direction. However, the synapse that connects it to another neuron only permits the action potential to move along the axon away from its own cell body.

Examination of the axons of various neurons with an electron microscope indicates that there are two different types of nerve fibers. The membranes of some axons are covered with a fatty insulating layer called *myelin* that has small uninsulated gaps called *nodes of Ranvier* every few millimeters (Fig. 9.1); these nerves are referred to as *myelinated nerves*. The axons of other nerves have no myelin sleeve (sheath), and these nerves are called *unmyelinated nerves*. This is a somewhat artificial classification; most human nerves have both types of fibers. Much of the early research on the electrical behavior of nerves was done on the unmyelinated giant nerve fiber of the squid (Fig. 9.4). Myelinated nerves, the most common type in humans, conduct action potentials much faster than unmyelinated nerves.

Two primary factors affect the speed of propagation of the action potential, the electrical resistance R within the core of the axon and the capacitance C (related to the charge stored) across the membrane. The time t needed to charge or discharge a simple series electrical circuit containing resistance and capacitance has an exponential functional form of exp (-t/RC) (see appendix B). The time constant τ is the value of τ when it equals RC, $\tau = t = RC$. A decrease in either R or C will decrease the time constant and the capacitor will charge or discharge faster.

The conduction speed of an action potential depends on the rate of charging or discharging an R-C circuit. The internal resistance of an axon, such as in Fig. 9.4, decreases as its diameter increases. For two axons with similar properties differing only in diameter, the larger diameter axon will have a faster conduction speed than an axon with a smaller diameter.

The greater the capacitance (or stored charge) of a membrane, the longer it takes to depolarize it, thus the slower the propagation speed. The myelin sleeve, as indicated in Fig. 9.1, is a good insulator, and this part of the axon has very low capacitance. Because of the low capacitance, the charge stored is very small compared to an unmyelinated section of a nerve with the same diameter and length. The conduction speed in myelinated fibers is much faster than in unmyelinated fibers. The unmyelinated squid axons (~1 mm in diameter) have propagation speeds of 20 to 50 m/s, whereas the myelinated fibers in man (about 10 μm in diameter) have propagation speeds of around 100 m/s. This large conduction speed results mainly from the very small capacitance of the myelinated axons.

Referring to Fig. 9.1, the action potential travels very fast in the myelinated portion and much slower in the unmyelinated sections (*nodes of Ranvier*). The action potential is reduced in amplitude in the myelinated segment, but restored to full size in the unmyelinated section that has conduction the same as in Fig. 9.4. Under these two conditions, the action potential travels very fast in the myelinated sections and much slower in the nodes; it thus appears to jump from one node of Ranvier to the next. This is called saltatory (leaping) conduction.

The advantage of myelinated nerves in man is their high propagation velocities in axons of small diameter. A large number of nerve fibers can thus be packed into a small bundle to provide many signal channels. For example, 10,000 myelinated fibers of 10 μm in diameter can be carried in a bundle with a cross-sectional area of 1 to 2 mm^2, whereas 10,000 unmyelinated fibers with the same conduction speed would be a bundle with a cross-sectional area of approximately 100 cm^2, or about 10,000 times larger.

9.2

PROBLEM

What is the advantage of myelinated nerves over unmyelinated nerves?

9.3

PROBLEM

What is the typical resting potential of a cell?

9.4

PROBLEM

What is the typical conduction velocity of the action potential in a nerve cell? What factors contribute to the conduction velocity?

9.3 Electrical Signals from Muscles—The Electromyogram

Diagnostic information about muscles can be obtained from their electrical activity. In this section we trace the transmission of the action potential from the axon into the muscle, where it causes muscle contraction. The record of the potentials from muscles during contraction is called the *electromyogram*, or EMG.

A muscle is made up of many motor units. A motor unit consists of a single branching neuron from the brain stem or spinal cord and the 25 to 2000 muscle fibers (cells) it connects to via motor end plates (Fig. 9.6a). The resting potential across the membrane of a muscle fiber is similar to the resting potential across a nerve fiber. Muscle action is initiated

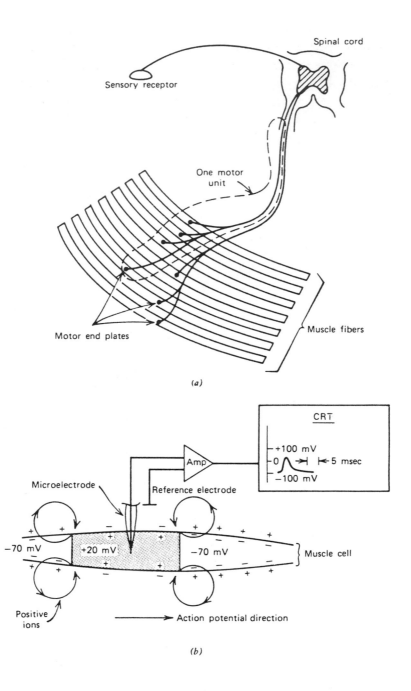

Figure 9.6. (a) Schematic of a neuron originating at the spinal cord and terminating on several muscle cells. The neuron and connecting muscle cells make up a motor unit (dashed line). (b) Instrument arrangement for measuring the action potential in a single muscle cell. The reference electrode is immersed in the fluid surrounding the cell.

by an action potential that travels along an axon and is transmitted across the motor end plates into the muscle fibers, causing them to contract. The record of the action potential in a single muscle cell is shown schematically in Fig. 9.6b. Such a measurement is made with a very tiny electrode (microelectrode) thrust through the muscle membrane.

EMG electrodes usually record the electrical activity from several fibers. Either a surface electrode or a concentric needle electrode is used. A surface electrode attached to the skin measures the electrical signals from many motor units. A concentric needle electrode inserted under the skin measures single motor unit activity by means of an insulated wire connected to its point. Figure 9.7 shows typical EMGs from the two types of electrodes. A typical arrangement for recording the EMG is shown in Fig. 9.8. The muscle's electrical signals can be displayed directly on one channel of an oscilloscope, and the signals can be

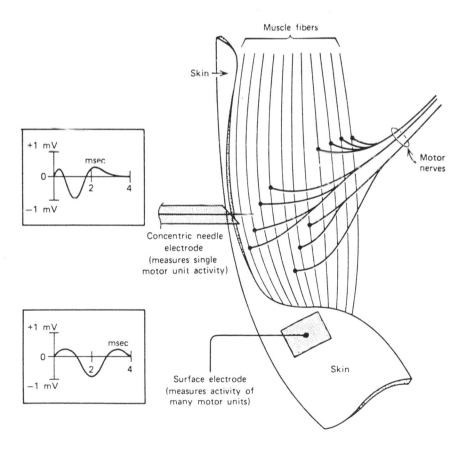

Figure 9.7. Electromyograms obtained with a concentric needle electrode and a surface electrode.

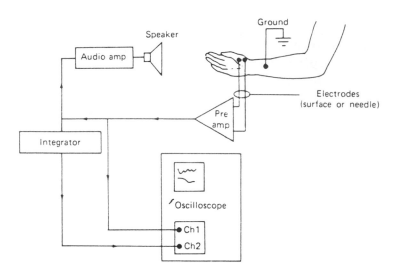

Figure 9.8. Instrument arrangement for obtaining an EMG.

integrated and displayed on a second channel. The signals can also be
passed through an amplifier and made audible by a loudspeaker. The
integrated record (in volt seconds) is a measure of the quantity of electricity associated with the muscle action potentials. Figure 9.9 shows
the EMG and its integrated form for different degrees of voluntary muscular contraction. More forceful contractions lead to greater action
potential activity. It is easier to evaluate the integrated form of action
potential activity because it is a smooth curve. In the clinic, the audible EMG and the integrated form are often used to determine the
condition of a muscle during contraction.

The EMG can be obtained from muscles or motor units that are stimulated electrically. This method is often preferred to voluntary
contraction. A voluntary contraction is usually spread over about 100
ms because all the motor units do not fire at the same time; also, each
motor unit may produce several action potentials depending upon the
signals sent from the central nervous system. With electrical stimulation, the stimulation time is well defined and all the muscle fibers fire
at nearly the same time. A typical stimulating pulse may have an amplitude of 100 V and last 0.1 to 0.5 ms.

An EMG obtained during electrical stimulation of a motor unit is
shown in Fig. 9.10. The action potential appears in the EMG after a
latency period (the time between stimulation and the beginning of the
response). Sometimes the EMGs from symmetrical muscles of the body

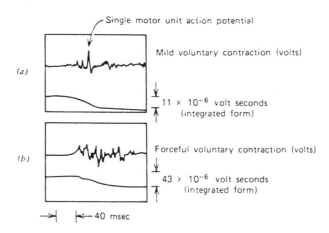

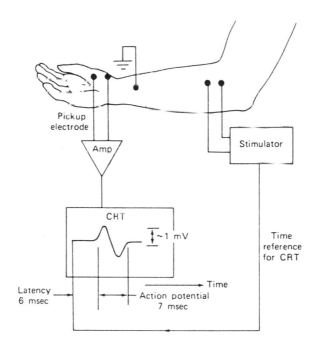

Figure 9.9. Electromyograms for (a) minimal contraction showing the action potential from a single motor unit and (b) maximal contraction showing the action potentials from many motor units. Note a and b have different scales. (Adapted from P. Strong, *Biophysical Measurements*, Tektronix, Inc., Beaverton, OR, 1970, p. 183, by permission of Tektronix, Inc. All rights reserved.)

Figure 9.10. Instrument arrangement for obtaining an EMG during electrical stimulation of a motor unit.

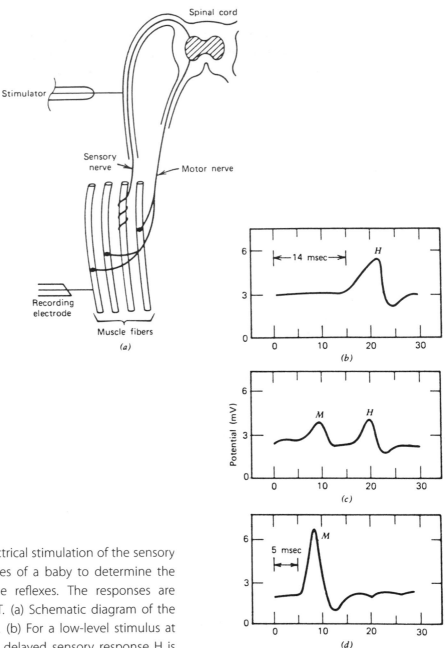

Figure 9.11. Electrical stimulation of the sensory and motor nerves of a baby to determine the condition of the reflexes. The responses are shown on a CRT. (a) Schematic diagram of the instrumentation. (b) For a low-level stimulus at time 0, a 14 ms delayed sensory response H is seen at the recording electrode. (c) For moderate stimuli two responses are obtained: the motor nerve M response at about 5 ms after the stimulus and the H response at 14 ms. (d) For a large stimulus only the M response is obtained. (Adapted from J. E. Thomas and E. H. Lambert, *J. Appl. Physiol.* 15: 1–9, 1960, Figure 7.)

are compared to each other or to those of normal individuals to determine whether the action potentials and latency periods are similar.

In addition to electrically stimulating the motor units, it is possible to excite the sensory nerves that carry information to the central nervous system. The reflex system can be studied by observing the reflex response at the muscle (Fig. 9.11). At low stimulating levels some of the sensitive sensory nerves are activated but the motor nerves are not and no M response is seen (Fig. 9.11b). The action potentials of the sensory nerves move to the spinal cord and generate the reflex response that travels along the motor nerves and initiates a delayed H response at the muscle. As the stimulus is increased, both the motor nerves and the sensory nerves are stimulated and both the M and the H responses are seen (Fig. 9.11c). At large stimulating levels only the M response is seen (Fig. 9.11d). The velocity of the action potential in motor nerves can also be determined. Stimuli are applied at two locations, and the latency period for each response is measured (Fig. 9.12). The difference between the two latency periods is the time required for an action potential to travel the distance

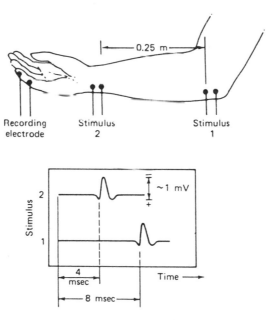

Figure 9.12. Method of measuring the motor nerve conduction velocity. The latency period for the response to stimulus 1 is 4 ms longer than that for the response to stimulus 2 ($\Delta t = 4 \times 10^{-3}$ s). The difference in distance Δx is 0.25 m; therefore, the nerve conduction velocity $v = \Delta x/\Delta t = 0.25$ m/4×10^{-3} s = 62.5 m/s.

between them; the velocity of the action potential is this distance divided by the difference in latency periods.

The conduction velocity for sensory nerves can be measured by stimulating at one site and recording at several locations that are known distances from the point of stimulation (Fig. 9.13). Many times nerve damage results in a decreased conduction velocity. Typical velocities are 40 to 60 m/s; a velocity below 10 m/s would indicate a problem. Electromyograms made during multiple stimulations are used to determine fatigue characteristics of muscles. The major muscles in humans can be re-stimulated at rates of between 5 and 15 Hz. Normal nerves and muscles show little change during prolonged re-stimulation as long as the rate of stimulation allows for a relaxation period of about 0.2 s between pulses. A patient with the relatively rare disease *myasthenia gravis* shows muscular weakness when carrying out a repetitive muscular task. The EMG of such a patient shows that in repetitive stimulation the muscle no longer responds.

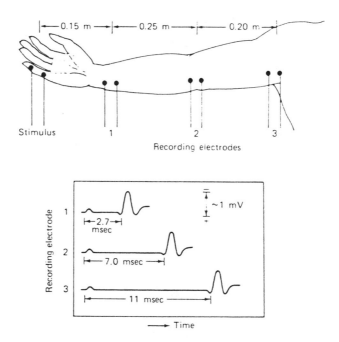

Figure 9.13. The sensory nerve conduction velocity can be determined by stimulating at one location and recording the responses with electrodes placed at known distances. The response traveled the 0.25 m from 1 to 2 in 4.3 ms; the conduction velocity is 0.25 m/4.3 × 10^{-3} s = 58 m/s. The conduction velocity from 2 to 3 is 0.20 m/4 × 10^{-3} s = 50 m/s.

9.4 Electrical Signals from the Heart—The Electrocardiogram

In Chapter 8, *Physics of the Cardiovascular System*, we discuss the heart as a double pump. It has four chambers (Fig. 9.14); the two upper chambers, the left and right atria, are synchronized to contract simultaneously, as are the two lower chambers, the left and right ventricles. The right atrium receives venous blood from the body and pumps it to the right ventricle. This ventricle pumps the blood through the lungs, where it is oxygenated. The blood then flows into the left atrium. The contraction of the left atrium moves the blood to the left ventricle, which contracts and pumps it into the general circulation; the blood passes through the capillaries into the venous system and returns to the right atrium.

The rhythmical action of the heart is controlled by an electrical signal initiated by spontaneous stimulation of special muscle cells located in the right atrium. These cells make up the *sinoatrial* (SA) *node*, or the *pacemaker* (Fig. 9.14). The SA node fires at regular intervals about 72 times per minute; however, the rate of firing can be increased or decreased by nerves external to the heart that respond to the blood demands of the body as well as to other stimuli. The electrical signal

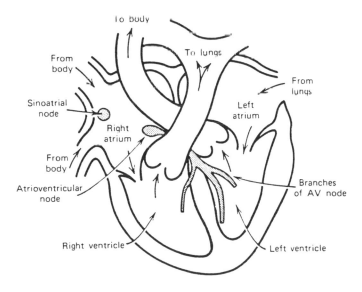

Figure 9.14. The human heart viewed from the front. Note the sinoatrial node, or the pacemaker, and the atrioventricular node, which initiates the contraction of the ventricles.

from the SA node initiates the depolarization of the muscle cells of both atria, causing the atria to contract and pump blood into the ventricles. Repolarization of the atria follows (see Fig. 9.5 for the shape of the action potential). The electrical signal then passes into the *atrioventricular* (AV) *node*, which initiates the depolarization of the right and left ventricles, causing them to contract and force blood into the pulmonary and general circulations. The ventricle muscles then repolarize and the sequence begins again.

The depolarization and repolarization of the heart muscle causes current to flow within the torso, causing electrical potentials on the skin. An electrocardiogram (ECG) is a recording of the electrical potentials between two points located at various positions on the surface of the body.

The relationship between the pumping action of the heart and the electrical potentials on the skin can be understood by considering the propagation of the depolarizing part of the heart's muscles, as shown in Fig. 9.15. See Fig. 9.5c for the shape of the action potential for heart muscle cells; these potentials exist much longer than the action potential in an axon; thus, it makes sense to look only at the depolarizing part of the heart in the following example. The resulting current flow in the torso leads to a potential drop as shown schematically on the resistor. The potential distribution for the entire heart when the ventricles are one-half depolarized is shown by the equipotential lines in Figure 9.16.

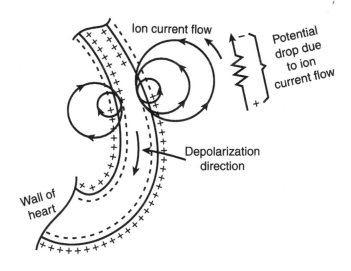

Figure 9.15. Schematic of the depolarizing wave moving down the heart wall generating ion currents indicated by the circles. The resistor suggests the electrical resistance of the tissues. The potential on the chest wall is due to current flow through the resistance of the torso.

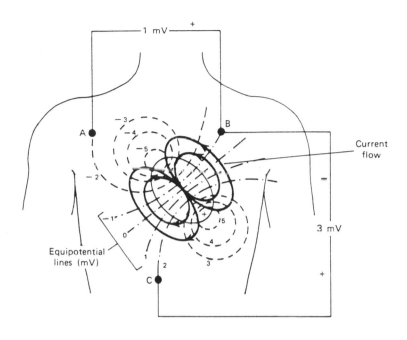

Figure 9.16. The potential distribution on the chest at the moment when the ventricles are one-half depolarized. Electrodes located at A, B, and C would indicate the potentials at that moment.

Note that the potentials measured on the surface of the body depend upon the location of the electrodes. The form of the potential lines shown in Fig. 9.16 is nearly the same as that obtained from an electric dipole. (An electric dipole is produced when equal positive and negative charges are separated from each other. It is represented by a vector.) In this case, it is the ion currents from the depolarizing heart that produce an electric *current* dipole. For this text, we call this a current dipole. The equipotential lines at other times in the heart's cycle can also be represented by a current dipole; however, the dipole changes in magnitude and orientation during the heart cycle. The current dipole model of the heart was first suggested by A. C. Waller in 1889 and has been modified by many others since.

The electrical (cardiac) potential that we measure on the body's surface is the instantaneous projection of the current dipole in a particular direction. As the vector changes with time, so does the projected potential. Fig. 9.17 shows a current dipole along with the three electrocardiographic body planes.

The surface electrodes for obtaining the ECG are most commonly located on the right arm (RA), the left arm (LA), and the left leg (LL), although the location of the electrodes can vary in different clinical

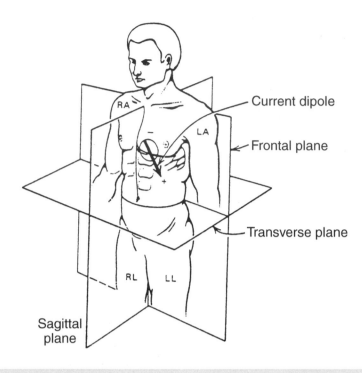

Figure 9.17. Electrocardiographic planes and a current dipole. RA, LA, RL, and LL indicate electrode locations on the right and left arms and legs.

situations; sometimes the hands or positions closer to the heart are used. In Fig. 9.18 these positions are labeled as potentials V_{RA}, V_{LA}, and V_{LL}, located at these respective positions. The measurement of the potential differences between the various combinations are called Lead I, Lead II, and Lead III measurements and written as:

$$\text{Lead I} = V_{LA} - V_{RA}$$

$$\text{Lead II} = V_{LL} - V_{RA} \qquad (9.1)$$

$$\text{Lead III} = V_{LL} - V_{LA}$$

This configuration was pioneered at the turn of the century by Willem Einthoven, a Dutch physiologist, and these three limb leads are called the *standard limb leads*. Usually, the three standard limb leads are used in a clinical examination. However, any two of the signals gives the relative amplitude and direction of the current dipole in the frontal plane (see Fig. 9.19).

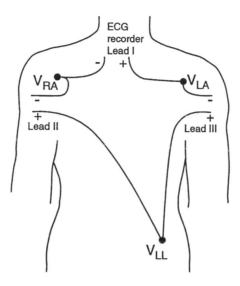

Figure 9.18. The potentials located at positions RA, LA, and LL are labeled V_{RA}, V_{LA}, and V_{LL}, respectively. The polarities of the recording instrument are shown where Lead I = $V_{LA} - V_{RA}$, Lead II = $V_{LL} - V_{RA}$, and Lead III = $V_{LL} - V_{LA}$, the standard limb measurements which are measured in the frontal plane, Fig. 9.17.

The augmented lead configurations, aV_R, aV_L, and aV_F, are also obtained in the frontal plane. For the aV_R lead configuration, an electric circuit combines the V_{LA} and V_{LL} potentials and takes their average $(V_{LA} + V_{LL})/2$. The difference between V_{RA} and this average gives aV_R. This provides a recording for a different projection of the current dipole signal as shown schematically in Fig. 9.20. In this case, the current dipole is shown by the + – signs and projects onto the line from V_{RA} to the center of the heart. The projection is shown below the line. In a similar fashion, the other augmented lead signals are obtained. They are described by:

$$aV_R = V_{RA} - (V_{LA} + V_{LL})/2$$

$$aV_L = V_{LA} - (V_{RA} + V_{LL})/2 \qquad (9.2)$$

$$aV_F = V_{LL} - (V_{RA} + V_{LA})/2$$

Taken together, the augmented aV_R configuration shown in Fig. 9.20 plus the aV_L and aV_F provide projections of the current dipole from

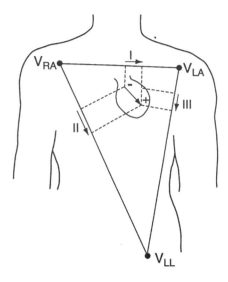

Figure 9.19. Schematic of the current dipole of the heart projected on the frontal plane. For electrical purposes the three electrodes located at the measuring positions of the potentials V_{RA}, V_{LA}, and V_{LL} can be thought of as the points of a triangle, the Einthoven triangle. The potential in Lead I at any moment is proportional to the projection of the current dipole on the line RA-LA; the potentials in Leads II and III are proportional to the projections on the other sides of the triangle.

approximately the heart to the right shoulder, left shoulder, and left leg, respectively. When these three signals are combined with the standard limb leads, the current dipole is overdetermined; however, the desired range of information is there for the physician to interpret.

Each ECG tracing maps out a projection of the current dipole, or the electrical activity of the heart, through each part of its cycle. Figure 9.21 shows schematically the Lead II output with the standard symbols for the parts of the pattern. The major electrical events of the normal heart cycle are: (1) the atrial depolarization, which produces the P wave; (2) the atrial repolarization, which is rarely seen and is unlabeled; (3) the ventricular depolarization, which produces the QRS complex; and (4) the ventricular repolarization, which produces the T wave. Figure 9.22 shows the six frontal plane ECGs for a normal subject. Note that in some cases the waveform is positive and in other cases it is negative; the sign of the waveform depends upon the direction of the current dipole and the polarity and position of the electrodes of the measuring instrument.

In a clinical examination, six transverse plane ECGs are usually made in addition to the six frontal plane ECGs. For the transverse plane

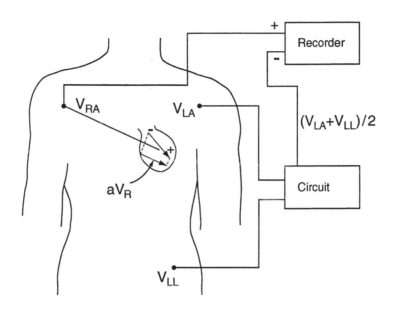

Figure 9.20. A schematic of the measurement of the augmented aV_R lead signal. The electronic circuit adds the V_{LA} and V_{LL} signals and provides the average $(V_{LA} + V_{LL})/2$. The aV_R is then the difference between the V_{RA} and the average. It represents approximately the projection of the current dipole, labeled by + –, in the direction of the heart to the right arm. The direction is shown by the line from V_{RA} to the center of the heart. The projection of the current dipole is below.

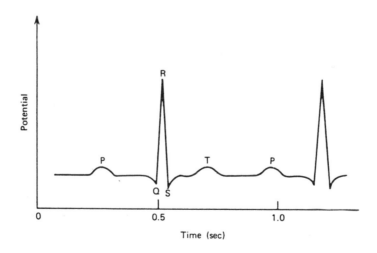

Figure 9.21. Typical ECG from Lead II position. P represents the atrial depolarization and contraction, the QRS complex indicates the ventricular depolarization, the ventricular contraction occurs between S and T, and T represents the ventricular repolarization.

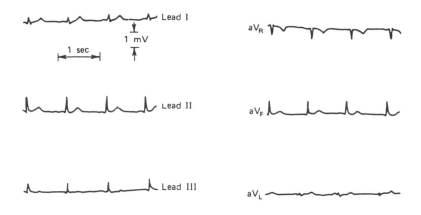

Figure 9.22. Six frontal place ECGs for a normal subject.

measurements the negative terminal of the recorder is attached to an *indifferent electrode*, which is an average obtained by the circuit from a combination of V_{RA}, V_{LA}, and V_{LL} (see Fig. 9.23a). The positive terminal is moved aross the chest wall to the locations labeled V_1 to V_6, as shown in Figures 9.23a and 9.23b.

The nine surface electrode locations produce 12 potentials: I, II, III, aV_R, aV_L, aV_F, and V_1 through V_6. Although this information overdetermines the current dipole behavior in the cardiac cycle, specialists have been trained with pattern recognition ability to assess quickly normal and abnormal heart conditions. Additional electrode positions are often used for research. ECGs are usually interpreted by cardiologists, who can quickly determine if the patterns are normal and if arrhythmias (rhythm disturbances) exist. However, computers are also used to analyze ECGs and do better than some cardiologists. In an intensive care unit (ICU) and during surgery, the ECG is continuously monitored and displayed on a video display terminal (oscilloscope).

An ECG shows disturbances in the normal electrical activity of the heart. For example, an ECG may signal the presence of an abnormal condition known as *heart block*. If the normal SA node signal is not conducted into the ventricle, then a pulse from the AV node will control the heartbeat at a frequency of 30 to 50 beats/min, which is much lower than normal (70 to 80 beats/min). While a heart block like this could make a patient a semi-invalid, an implanted pacemaker enables the individual to live a reasonably normal life.

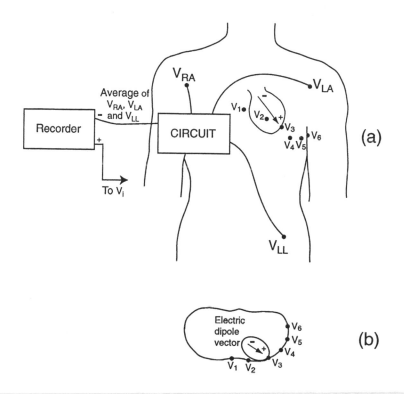

Figure 9.23. Transverse plane ECG positions. (a) Frontal view showing an electronic circuit that provides an indifferent (negative) electrode obtained by averaging V_{RA}, V_{LA}, and V_{LL}. The positive electrode is moved to the chest positions labeled by the various V_1 through V_6 locations. (b) Top view.

9.5

PROBLEM

What important role is performed by the SA node of the heart?

9.6

PROBLEM

Give the locations of the electrodes for the standard ECG limb leads.

PROBLEM 9.7

Sketch an augmented ECG lead. Why are these used?

PROBLEM 9.8

What electrical phenomenon in the heart produces the QRS complex of the ECG?

9.5 Electrical Signals from the Brain—The Electroencephalogram

If you place electrodes on the scalp and measure the electrical activity, you will obtain some very weak complex electrical signals. These signals are due primarily to an electrical activity of the neurons in the cortex of the brain. They were first observed by Hans Berger in 1929; since then much research has been done on clinical, physiological, and psychological applications of these signals, but a basic understanding is still lacking. One hypothesis is that the potentials are produced through an intermittent synchronization process involving the neurons in the cortex, with different groups of neurons becoming synchronized at different instants of time. According to this hypothesis the signals from points on the right side are compared to signals from symmetrical points on the left side.

The recording of the signals from the brain is called the *electroencephalogram* (EEG). Electrodes for recording the signals are often small discs of chlorided silver. They are attached to the head at locations that depend upon the part of the brain to be studied. Figure 9.24 shows the international standard 10-20 system of electrode location, and Figure 9.25 shows typical EEGs for several pairs of electrodes. The reference electrode is usually attached to the ear (A_1 or A_2 in Fig. 9.24). In routine

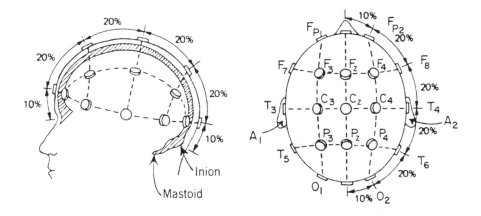

Figure 9.24. International standard 10-20 system of electrode location for EEGs. Lettered electrodes are located at intervals of 10% and 20% of the distances between specific points on the skull. The inion is the bony protuberance at the lower back of the skull and the mastoid is that behind the ear.

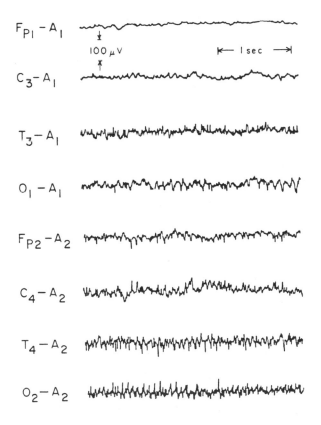

Figure 9.25. Normal EEGs. See Fig. 9.24 for the location of the electrodes. The reference electrode is connected to the ear (A_1 or A_2).

exams, 8 to 16 channels are recorded simultaneously. Since asymmetrical activity is often an indication of brain disease, the right side signals are often compared to the left side signals.

The international 10-20 system has 21 surface electrodes. Sometimes additional intermediate electrodes located between the 10-20 system electrodes are also used. The results provide more information than the 10-20 spacing.

The amplitude of the EEG signals is low (about 50 μV), and interference from external electrical signals often cause artifacts in the EEG. Even if the external noise is eliminated, extraneous potentials from muscle activity, such as eye movement, can cause artifacts in the record. The frequencies of the EEG signals are dependent upon the mental activity of the subject. For example, a relaxed person usually has an EEG signal composed primarily of frequencies from 8 to 13 Hz, or *alpha waves*. When a person is more alert, a higher frequency range, the *beta*

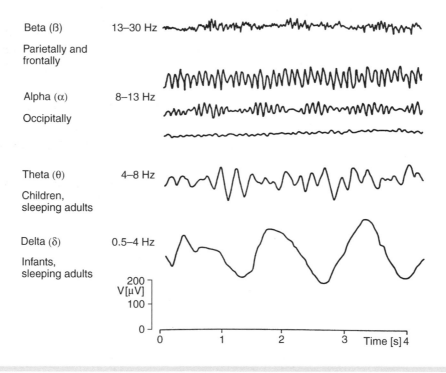

Figure 9.26. Examples of EEG waves. (From BIOELECTROMAGNETISM, PRINCIPLES AND APPLICATIONS OF BIOELECTRIC AND BIOMAGNETIC FIELDS by Jaakko Malmivuo and Robert Plonsey. Copyright © 1995 by Oxford University Press, Inc. Used by permission of Oxford University Press, Inc.)

wave range (above 13 Hz), dominates the EEG signal (see Fig. 9.26 for EEG wave examples). The various frequency bands are as follows:

Delta (Δ), or slow	0.5 to 3.5 Hz
Theta (θ), or intermediate slow	4 to 7 Hz
Alpha (α)	8 to 13 Hz
Beta (β), or fast	>13 H

The EEG is used as an aid in the diagnosis of diseases involving the brain. It is most useful in the diagnosis of epilepsy and allows classification of epileptic seizures. The EEG for a severe epileptic attack with loss of consciousness, called a *grand mal* seizure, shows fast high voltage spikes in all leads from the skull (Fig. 9.27a). The EEG for a less severe attack, called a *petit mal* seizure, shows up to 3 rounded waves per second followed or preceded by fast spikes (Fig. 9.27b).

The EEG aids in the diagnosis of brain tumors since electrical activity is reduced in the region of a tumor. Other more quantitative methods for locating brain tumors involve computerized tomography (CT) or magnetic resonance imaging (MRI).

The EEG is used as a monitor in surgery when the ECG cannot be used. It is also useful in surgery for indicating the anesthesia level of the patient. During surgery a single pair of EEG electrodes is usually monitored.

Much sleep research involves observing the EEG patterns for various stages of sleep (Fig. 9.28). As a person becomes drowsy, particularly

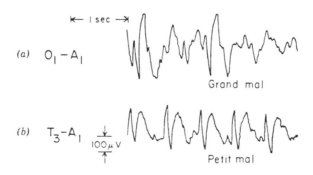

Figure 9.27. Electroencephalograms for two types of epilepsy: (a) grand mal and (b) petit mal.

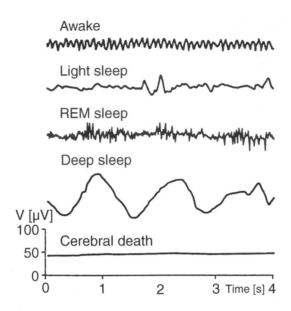

Figure 9.28. Examples of EEG signals for different levels of sleep, including eternal sleep (death). (From BIOELECTROMAGNETISM, PRINCIPLES AND APPLICATIONS OF BIOMAGNETIC FIELDS by Jaakko Malmivuo and Robert Plonsey. Copyright © 1995 by Oxford University Press, Inc. Used by permission of Oxford University Press, Inc.)

with the eyes closed, alpha waves (8 to 13 Hz) dominate the EEG. The amplitude increases and the frequency decreases as a person moves from light sleep to deeper sleep. Occasionally an EEG taken during sleep shows a high frequency pattern called *paradoxical sleep* or *rapid eye movement* (REM) *sleep* because the eyes move during this period. Paradoxical sleep appears to be associated with dreaming.

Besides recording the spontaneous activity of the brain, we can record EEG signals when the brain receives external stimuli such as flashing lights or pulses of sound. Signals of this type are called *evoked responses*. Fig. 9.29a shows three EEGs taken during the early stages of sleep with a series of 10 sound pulses used as an external stimulus. The EEGs show responses to the first few pulses and the last two pulses. The lack of responses in between is called *habituation*.

Because the evoked response is small and may be lost in the variable EEG signal, the signal is repeated and the EEG responses are averaged by computer. Random signals from the normal EEG tend to average to a background signal and the evoked response becomes clear. Fig. 9.29b shows an evoked response averaged for 64 stimuli.

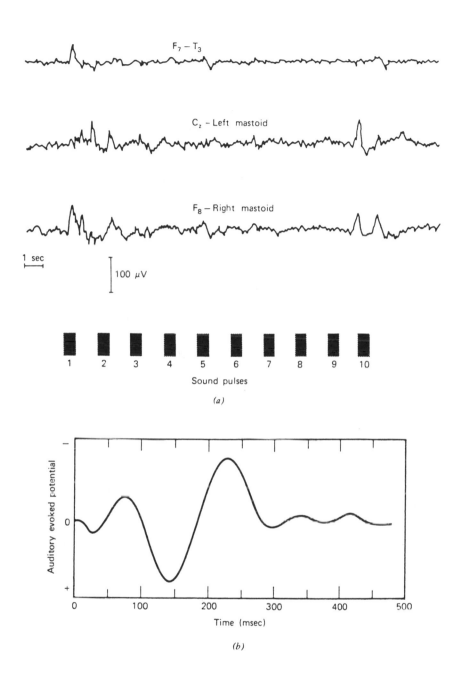

Figure 9.29. (a) Several EEGs taken during early sleep with 2 Hz sound pulses used as stimuli. (b) Evoked response averaged for 64 sound stimuli. (Courtesy of Dr. Lloyd F. Elfner, Director, and D. Gustafson, Psychoacoustics Laboratory, Florida State University, Tallahassee, FL.)

9.6 Electrical Signals from the Eye—The Electroretinogram and the Electrooculogram

The recording of potential changes produced by the eye when the retina is exposed to a flash of light is called the *electroretinogram* (ERG). One electrode is located in a contact lens that fits over the cornea and the other electrode is attached to the ear or forehead to approximate the potential at the back of the eye (Fig. 9.30).

An ERG signal is more complicated than a nerve axon signal because it is the sum of many effects taking place within the eye. The general form of an ERG is shown in Fig. 9.31. The B wave is the most interesting clinically since it arises in the retina. The B wave is absent

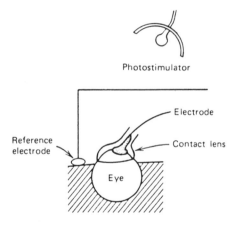

Figure 9.30. The placement of electrodes for obtaining an ERG. The reference electrode is on an ear or the forehead.

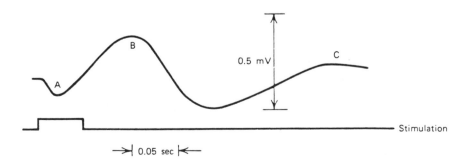

Figure 9.31. A smoothed version of an ERG. The letters identify portions of a normal ERG.

in the ERG of a patient with inflammation of the retina that results in pigment changes, or *retinitis pigmentosa*.

Not shown in Fig. 9.31 is the additional detail coming mostly from various regions in the retina. While the ERG has not played a major role in clinical applications, two Nobel Prizes, in 1967 and 1981, were awarded for studies involving the ERG.

The *electrooculogram* (EOG) is a recording of potential changes due to eye movement. For this measurement, a pair of electrodes is attached near each side of the eye (Fig. 9.32a). The EOG potential is defined as zero with the eye looking straight ahead at the reference spot $0°$. Figure 9.32b shows the EOG potential change for horizontal movement of the eyeball.

The EOG in Fig. 9.32b provides information on the orientation of the eye; however, the angular velocity and its angular acceleration can also be determined. Electrooculography is primarily a research technique. EOG studies have been done to determine the effects of drugs on eye movement, eye movement during sleep, and during visual search. One clinical application is to study eye movement during *nystagmus*, a condition concerned with small movements of the eye. The EOG signals depend on both the eye muscle and the vestibular (balance) system.

9.9

PROBLEM

What are alpha waves on an EEG? When would a person normally have alpha waves on an EEG?

9.10

PROBLEM

What is an evoked EEG response?

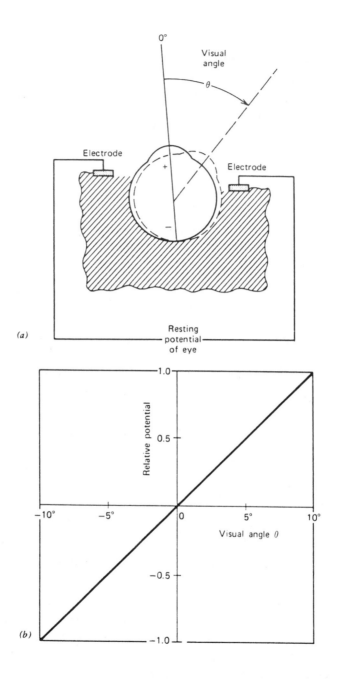

Figure 9.32. (a) For obtaining the EOG an electrode is mounted on each side of the eye. The visual angle is indicated. (b) The change of potential is plotted as a function of visual angle.

What is meant by REM sleep?

9.7 Magnetic Signals from the Heart and Brain—The Magnetocardiogram and the Magnetoencephalogram

Since a flow of electrical charge produces a magnetic field, the current in the heart during depolarization and repolarization also produces a magnetic field. Magnetocardiography measures these very weak magnetic fields around the heart. The recording of the heart's magnetic field is the *magnetocardiogram* (MCG).

The magnetic field around the heart is about 5×10^{-11} tesla (T), or about one-millionth the strength of the earth's magnetic field. [The centimeter-gram-second (cgs) unit for magnetic fields is the gauss; 1 T $= 10^4$ gauss.] To measure these very weak fields it is necessary to use magnetically shielded rooms and very sensitive magnetic field detectors (magnetometers). One such detector, called a SQUID (Superconducting QUantum Interference Device), operates at a temperature of about 5 K and can detect both steady and alternating magnetic fields as small as 10^{-14} T.

The SQUID magnetometer has also been used to record the magnetic field surrounding the brain. The recording of this field is called the *magnetoencephalogram* (MEG). During the alpha rhythm, the magnetic field from the brain is about 1×10^{-13} T. This is almost one-billionth of the earth's magnetic field.

Neither the MCG nor the MEG is likely to show up soon in routine medical measurements because of the substantial instrument cost, the increased technical difficulty to make and interpret the measurements, and the question of its diagnostic importance is still being established. However, the MEG particularly shows considerable promise for brain studies and is worth discussing in more detail. While with the EEG it is very difficult to localize where a particular signal originates in the brain, with the MEG, using evoked stimulation by sound, light, touch, smell or external pulsed magnetic fields, the particular stimulation can be repeated and the magnetic responses stored in a computer. This provides an enhancement of the small desired signals while averaging out both undesired MEG signals from other parts of the brain and signals from magnetic noise. With magnetic signals measured over a significant region of the brain, the location in the brain where the MEG signals originate can be determined in terms of a current dipole. Recall the current dipole model for the heart in Figure 9.16. It changed magnitude and direction with time as the heart went through its cycle. In a similar fashion one finds the current dipole from stimulation of the brain.

In Fig. 9.33, a modern detector arrangement used to measure the MEG is shown. In the example, the subject is in a magnetically shielded

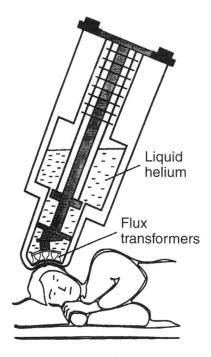

Figure 9.33. Detector arrangement for measuring the magnetoencephalogram (MEG). The liquid helium-cooled dewar contains the SQUID magnetometer sensors (called flux transformers) which are arranged to detect the perpendicular to the surface magnetic fields very close to the skull. In this case, the patient's brain can be stimulated by an audible signal via plastic tubes and an earpiece. (Used by permission from "Magnetoencephalography—Theory, Instrumentation, and Applications to Noninvasive Studies of the Working Human Brain." Matti Hämäläinen et al., *Reviews of Modern Physics*, Vol. 65, No. 2, 1993, p. 419.)

room, on a bed with a pillow to fix the head and prevent movement. There are multiple SQUID magnetometers (called flux transformers) arranged at the bottom of a liquid helium cooled (5K) dewar which are brought as close to the head as possible. Usually the perpendicular component of the magnetic field is measured.

Figure 9.34 shows the measured magnetic field signals obtained as a function of time after an auditory evoked response. Each of the scans shows the average of 66 stimuli averaged measurements to illustrate the reproducibility of the measurements. The time of the stimuli corresponds to the start of each of the scans. The magnetic field measurements correspond to the change in magnetic field components in the x (upper scan) and y directions (lower scan) for each of the 12

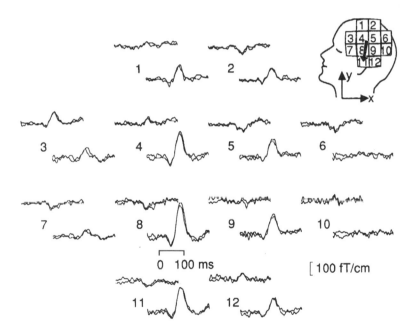

Figure 9.34. Measured stimulated auditory responses for the magnetic field measured at 12 locations as shown by the inset, upper right. For each location there are two recordings averaged over 66 evoked auditory stimulations. For each, the upper and lower recordings refer to the change in the perpendicular magnetic field (z direction) with directions y and x, respectively. The stimulus began at t = 0, the start of the recording and the time scale is given below the data for the region labeled 8. The magnetic field scale is to the right of the 8 label and is given in fT/cm ($f = 10^{-15}$). (Used by permission from "Magnetoencephalography—Theory, Instrumentation, and Applications to Noninvasive Studies of the Working Human Brain." Matti Hämäläinen et al., *Reviews of Modern Physics,* Vol. 65, No. 2, 1993, p. 420.)

measurement locations shown in the inset. The equivalent current dipole, shown in the inset by an arrow, is determined for the time corresponding to the largest peak in the measurements. Generally, the current dipole would be evaluated for all times of the measurements to provide its magnitude, direction, and location in the brain.

Not all magnetic fields produced within the body are the result of ion currents; the body can be easily contaminated with magnetic materials. For example, asbestos workers inhale asbestos fibers, which contain iron oxide particles. The size of the magnetic field from the iron oxide in a worker's lungs can be used to estimate the amount of inhaled asbestos dust. Typical magnetic fields from asbestos workers' chests are about one-thousandth of the earth's magnetic field (5×10^{-8} T). Figure 9.35 shows scans of persons with magnetic contamination.

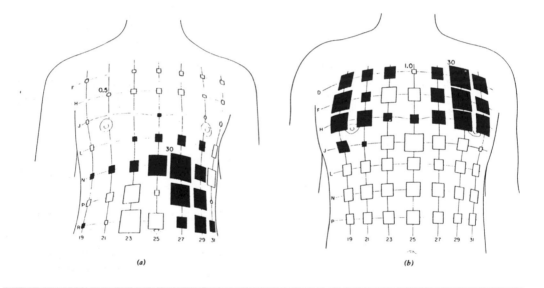

Figure 9.35. The magnetic field distributions from the chests of two subjects with magnetic contamination. The black squares show the fields in one direction while the open squares show the return paths. The area of each square is proportional to the strength of the field; values of the largest and smallest magnetic fields are given in units of 10^{-11} T. these patterns were due to (a) about 100 mg of iron oxide in the lungs of a welder. (From D. Cohen, *Science*, 180 (4087), p. 747. Copyright 1973 by the American Association for the Advancement of Science.)

9.8 Current Research Involving Electricity in the Body

Many electrical phenomena exist in the body, and our understanding of them varies widely. In this section some of the phenomena currently being explored are discussed. As our knowledge about electricity in the body increases, we should find more ways to use it in the diagnosis and treatment of diseases.

One life process that appears to be electrically controlled is bone growth. Bone contains collagen, which is a piezoelectric material; when a force is applied to collagen, a small dc electrical potential is generated. The collagen conducts current mainly by negative charges. Mineral crystals of the bone (apatite) close to the collagen conduct current by positive charges. At a junction of these two types of semiconductors, current flows easily in one direction but not in the other direction. (This is the basic idea of changing an ac signal to a dc signal by rectification.) It is thought that the forces on bones produce potentials by the piezoelectric effect and that at the junctions of collagen-apatite, currents are produced that induce and control bone growth. The currents are proportional to stress (force per unit area), so increased mechanical bone stress results in increased growth.

Another small direct current arises in an injured zone and is called the *injury current*. The electrical potential at a site of injury is higher than that in surrounding areas. This high potential is believed to be associated with limb regeneration in animals like the salamander and with fracture and wound healing in man. Stimulation of fracture sites with a direct current of 1 to 3 nA has been found to promote healing of bone fractures and bone conditions involving poor growth. It also enhances healing of burned areas.

Although the autonomic nervous system is not generally under voluntary control, it can be influenced by external stimuli. One means of influencing the system that has been known for some time is called *biofeedback*. Recently there has been renewed interest in biofeedback. While early research on biofeedback has been promising, many aspects are still not understood. As we learn more about it, we may be able to utilize it more in medicine.

If we want to use feedback to control the output of some device, whether an amplifier or some part of the body, we first measure the output to see what is happening. We then feed back this information to the input to affect the output in a desired manner. Negative feedback produces a stable output and is involved with the regulation of many body functions. We see negative feedback in the decrease in the diameter of the iris when a person is subjected to a bright light. The bright light increases the optic nerve signal to the brain, and the brain in turn decreases the diameter of the iris, thus decreasing the optic nerve signal. Other examples are the increase in heart rate and breathing with exercise and the use of shivering and perspiring to control body temperature.

In biofeedback the individual is consciously part of the feedback circuit. Sensors that monitor a subject's skin temperature, brain signals, or nerve action provide signals that are amplified and presented to the subject, who then tries through concentration to obtain a desired effect.

Through biofeedback body functions that are normally controlled by the autonomic nervous system can be consciously controlled. For example, EEG studies have shown that the alpha rhythm (8 to 13 Hz) indicates a low-arousal or relaxed state of the body, a condition that is often sought in biofeedback studies. If a subject finds that his/her EEG output changes from alpha rhythm to beta activity (greater than 13 Hz) when a headache is developing, the subject can, by mental relaxation, persuade both brain and body to return to alpha rhythm, thus forestalling the headache. Muscle relaxation can also be achieved through biofeedback; in this case, the EMG from a tense muscle is the signal presented to the patient. In addition, biofeedback has been used to control high blood pressure, acidity levels in the stomach, and irregular heart activity.

10

Sound and Speech

In this chapter we discuss the physical processes of sound, the importance of sound in diagnostic medicine through such procedures as percussion and auscultation, and the production of sound by the voice (phonation).

Medical specialists concerned with the ear and hearing are: (1) the *otolaryngologist*, an M.D. who specializes in diseases of the ear and throat; (2) the *otorhinolaryngologist*, an M.D. who specializes in diseases of the ear, nose, and throat (also called an ENT specialist); and (3) the *audiologist*, a non-M.D. who specializes in measuring hearing response, diagnosing hearing disorders through hearing tests, and rehabilitating those with varying degrees of hearing loss. (See Chapter 11, *Physics of the Ear and Hearing*.)

Sound is a major method of communication and gives us enjoyment in the form of music. However, noise pollution, or noise at undesirable levels, is an ever present problem in modern society. Federal regulations now limit the noise caused by cars and trucks on the highways (permissible noise levels) and regulate sound levels in the workplace. Noise levels in the workplace are monitored by the Occupational Safety and Health Administration (OSHA). These limits are fixed at 85 decibels (dB) for 8 hours exposure before hearing protection must be used. (The *decibel*, the unit of sound level intensity, is discussed below.)

Infrasound refers to sound frequencies below the normal hearing range, or less than 20 hertz. (1 hertz (Hz) = 1 s^{-1}). Infrasound is produced by natural phenomena like earthquake waves and atmospheric pressure changes; it can also be produced mechanically, such as by a blower in a ventilator system or an open window in a moving car. A typical ventilator system produces frequencies of about 10 Hz. These frequencies cannot be heard, but they can cause headaches and physiological disturbances.

The audible sound range is usually defined as 20 Hz to 20,000 Hz (20 kHz). However, relatively few people can hear over this entire range. Older people often lose the ability to hear the frequencies above 10 kHz. The frequency range above 20 kHz is called *ultrasound*. (Ultrasound should not be confused with supersonic, which refers to velocities faster than the velocity of sound in a medium.) Ultrasound is used clinically in a number of specialties. There is a growing trend to locate ultrasound equipment in the diagnostic radiology area, and some diagnostic radiologists specialize in ultrasonic imaging of the body. Ultrasound is also used by obstetricians to examine the unborn child. It often gives more information than an x-ray. It is generally believed that ultrasound is less hazardous to the fetus, although there is no evidence to suggest a measurable hazard from either modality.

10.1 General Properties of Sound

10.1.1 Physical Properties of Sound

A sound wave is a mechanical vibration in a gas, liquid, or solid that travels outward from the source with some definite velocity. We show schematically in Fig. 10.1a the diaphragm of a loudspeaker vibrating back and forth in air at a frequency *f*. This is an example of the production of a sound wave. The vibrations cause local increases and decreases in pressure relative to atmospheric pressure (Fig.10.1b). These pressure increases, called *compressions*, and decreases, called *rarefactions*, spread outward as a longitudinal wave; that is, a wave in which the air molecules move back and forth in the same direction that the wave travels. The compressions and rarefactions can also be described by density changes and by displacement of the atoms and molecules from their equilibrium positions.

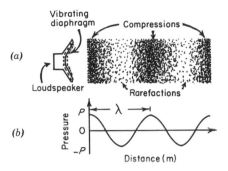

Figure 10.1. Schematic representation of a sound wave from a loudspeaker frozen in time at the peak of compression. (a) A diaphragm vibrates at a frequency *f* and produces compressions (increased pressure) and rarefactions (decreased pressure) in air. (b) The pressure of the sound wave relative to atmospheric pressure versus distance. P is the maximum pressure variation from atmospheric, and λ is the wavelength.

The relationship between the frequency of vibration f of the sound wave, the wavelength λ, and the velocity v of the sound wave is

$$v = \lambda f \quad \text{or} \quad \lambda = \frac{v}{f} \tag{10.1}$$

For example, using the second form, we can calculate the wavelength in air at 20 C knowing that v = 344 m/s and f = 1000 Hz as follows:

$$\lambda = \frac{344 \text{ ms}^{-1}}{1000 \text{ s}^{-1}} = 0.344 \text{ m.}$$

Energy is carried by the wave as both potential and kinetic energy. The intensity I of a sound wave is defined as the energy per second (1 J/s = 1 watt) carried by the sound wave through a cross-sectional area of one m^2 [unit: watts/m^2]. For a plane wave, I is given by

$$I = 1/2 \; \rho v A^2 (2\pi f)^2 = 1/2 \; Z(A\omega)^2 \tag{10.2}$$

where ρ is the density of the medium; v is the velocity of sound; f is the frequency; $\omega = 2\pi f$ is the angular frequency in rad/s; A is the maximum displacement amplitude of the atoms or molecules from the equilibrium position; and $Z = \rho v$ is the acoustic impedance. The acoustic impedance is important when calculating reflections (echoes) or transmission of sound when it strikes a barrier or a medium where the velocity of sound changes. Some typical values of ρ, v, and Z are given in Table 10.1. The intensity can also be expressed as

$$I = \frac{P^2}{2Z} \qquad (10.3)$$

where P is the maximum change from atmospheric pressure (see Example 10.1).

Table 10.1. Values of ρ, v, and Z for Various Substances

	ρ (kg/m^3)	v (m/s)	Z (kg/m$^2\cdot$s)
Air	1.29	3.31×10^2	4.30×10^2
Water	1.00×10^3	14.8×10^2	1.48×10^6
Fat	0.92×10^3	14.5×10^2	1.33×10^6
Muscle	1.04×10^3	15.8×10^2	1.64×10^6

10.1

PROBLEM

What is the range of wavelengths of sound waves for the range of frequencies that can be heard by the human ear, $f = 20$ Hz to 20,000 Hz? (Take the velocity of sound in air at 20 C to be $v = 344$ m/s.) [Answer: $\lambda = 17.2$ m to 1.72×10^{-2} m (1.72 cm)]

10.2

PROBLEM

What is the wavelength of a 1000 Hz sound wave in water if its velocity in water is $v = 1480$ m/s? [Answer: $\lambda = 1.48$ m]

10.3

PROBLEM

The maximum sound intensity that the ear can tolerate at 1000 Hz is approximately $I = 1$ W/m^2. What is the maximum displacement amplitude, A, in air corresponding to this intensity? (Note that the displacement you find here is approximately the diameter of a single cell!) [Answer: $A = 1.1 \times 10^{-5}$ m]

10.4

PROBLEM

The faintest sound intensity the ear can detect at 1000 Hz is approx-imately 10^{-12} W/m^2. What is A under these conditions? (Here the displacement you find is smaller than the diameter of the hydrogen atom!) [Answer: A = 1.1×10^{-11} m]

10.5

PROBLEM

What is the approximate range of movement of the eardrum at the threshold of hearing at a frequency of $f = 3000$ Hz?
[Answer: A = 3.7×10^{-12} m]

10.1.2 Sound Intensity Level—The Decibel (dB)

Often it is the comparison of two sound intensities and not their mag-nitudes that is important. A special unit, named the *bel*, has been developed for the sound intensity level which compares the ratio of two sound intensities. This unit was named after Alexander Graham Bell, who invented the telephone and did research in sound and hearing. The bel is defined by the expression $\log_{10}(I_2/I_1)$, where I_2/I_1 is the ratio of two sound intensities. Thus, if one sound is 10 times more intense than the other, $I_2/I_1 = 10$, then $\log_{10}(10) = 1.0$ or the two sound intensities differ by 1 bel. Because the bel is a rather large unit, it is common to use a tenth of a bel or *decibel* (dB); thus, 1 bel = 10 dB.

The sound intensity level, L_I, which is in common use is defined by the equation

$$L_I \text{ (dB)} = 10 \log_{10} (I_2/I_1) \tag{10.4}$$

where the (dimensionless) unit of measure is the *decibel*, or dB. [See Appendix C for a review of the logarithm function.] Similarly, the sound pressure level, L_P, is defined by the equation

$$L_P \text{ (dB)} = 20 \log_{10} (P_2/P_1) \tag{10.5}$$

which results from the property of logarithms that, for any x, $\log_{10}(x^2)$ = 2 $\log_{10}(x)$, and from equation (10.3) which shows that I is proportional to P^2.

As an example, equation (10.5) can be used to show that for two sounds with pressures that differ by a factor of 2 ($P_2/P_1 = 2$),

$$L_P = 20 \log_{10}(P_2/P_1) = 20 \log_{10}(2) = 20(0.301) = 6 \text{ dB}.$$

Thus, a hi-fi set that gives a sound output uniform to $\pm$ 3 dB (a total variation of 6 dB) from 30 to 15,000 Hz has a sound pressure variation over its frequency range of no more than $P_2/P_1 = 2$. This variation would not be noticed by the average ear except under controlled laboratory conditions.

For hearing tests, it is convenient to use a reference sound intensity (or sound pressure) to which other sound intensities can be compared. The reference sound intensity $I_0 = 10^{-12}$ W/m², which corresponds to a reference pressure level $P_0 \cong 2 \times 10^{-5}$ Pa. A 1000 Hz note of this intensity is barely audible to a person with good hearing.

If a sound intensity is given in dB with no reference to any other sound intensity, you can assume that I_0 is the reference intensity. Table 10.2 gives the intensities of some typical sounds in terms of this reference value. The most intense sound that the ear can tolerate without pain is about $L_I = 120$ dB. This corresponds to $I/I_0 = 10^{12}$ and $P/P_0 = 10^6$.

Table 10.2. Intensities (Approximate) of Various Sounds

	Intensity (W/m²)	L_I (dB)
Threshold of perception	10^{-12}	0
Whisper	10^{-10}	20
Average dwelling	10^{-9}	30
Business office	10^{-7}	50
Speech at 1 m	10^{-6}	60
Busy street	10^{-5}	70
Subway or automobile	10^{-3}	90
Sound that produces pain	10^{0}	120
Jet aircraft on takeoff	10^{1}	130
Rocket on launch pad	10^{5}	170

10.6

PROBLEM

What is the maximum displacement, in air, for a 1000 Hz sound wave with intensity of $L_I = 50$ dB?
[Answer: $A = 3.4 \times 10^{-9}$ m]

10.7

PROBLEM

What is the sound intensity level, L_I, of your loudest shout relative to your normal speaking voice if their intensities are in the ratio of $10^3:1$? From Table 10.2, what is the approximate sound intensity level of this shout relative to I_0? (Assume you are measuring it at a distance of 1 meter from you.)
[Answer: 30 dB; $I = 10^{-9}$ W/m^2]

10.8

PROBLEM

(This problem is meant to help you explore the properties of the logarithm function.)
(a) By what factor does the intensity ratio, I/I_0, change for each 3 dB change in the sound intensity level, L_I?
  [Answer: By a factor of 2; $I/I_0 = 2$]
(b) By what factor does the intensity ratio, I/I_0, change for each 6 dB change in the sound intensity level, L_I?
  [Answer: By a factor of 4; $I/I_0 = 4$]
(c) By what factor does the intensity ratio, I/I_0, change for each 10 dB change in the sound intensity level, L_I?
  [Answer: By a factor of 10; $I/I_0 = 10^1$]

10.1.3 Reflection and Transmission of Sound at Barriers

Echoes

When a sound wave hits the body, or for that matter, any region where the velocity of sound changes, part of the wave is reflected and part is transmitted (Fig. 10.2). The ratio of the reflected pressure amplitude R

to the incident pressure amplitude A_0 depends on the acoustic impedances of the two media, Z_1 and Z_2. The relationship is

$$R/A_0 = (Z_2 - Z_1)/(Z_1 + Z_2) \qquad (10.6)$$

For a sound wave in air hitting the body, Z_1 is the acoustic impedance of air and Z_2 is the acoustic impedance of tissue. For a sound wave
going from one medium to another within the body (e.g., from muscle
to bone, from fat to muscle, or from any tissue to an air space), Z_1 is
the acoustic impedance of the first region, and Z_2 is that of the second.
Note that if $Z_1 = Z_2$, there is no reflected wave, and transmission to the
second medium is complete. Also, if $Z_2 < Z_1$, the sign change indicates
a 180° phase change of the reflected wave.

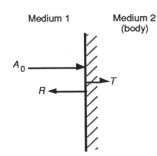

Figure 10.2. A sound wave of pressure amplitude A_0
incident upon the body. Part of the wave (of pressure
amplitude R) is reflected, and part (of pressure amplitude T)
is transmitted.

Transmitted Sound

The ratio of the transmitted pressure amplitude T to the incident wave
amplitude A_0 is

$$T/A_0 = 2Z_2/(Z_1 + Z_2) \qquad (10.7)$$

(Equations 10.6 and 10.7 are for sound waves striking perpendicular to
the surface. The equations for sound waves hitting from various angles
are more complicated and are not given here. As you might expect, they
are similar to the equations which occur in the refraction of light waves,
including Snell's law.)

While these equations give the ratio of the pressure amplitudes, it
is the reflected and transmitted intensities which are measured.

10.9

PROBLEM

From equations 10.2, 10.6, and 10.7 show that the ratio of the reflected and transmitted intensities are, respectively, $(R/A_0)^2$ and $(Z_1/Z_2)(T/A_0)^2$.

A good example of a large reflection of sound is when you are standing at the side of a swimming pool trying to talk with a friend who is under water. Most of the sound energy is reflected at the surface of the water, and only a small part is transmitted. We describe this by saying that the impedance mismatch between these two materials (air and water) is large; a calculation shows that nearly 99.9% of the incident intensity is reflected and 0.1% is transmitted (see Example 10.1). This corresponds to a 30 dB decrease in transmitted signal from air to water. Whenever acoustic impedances differ greatly, there is almost complete reflection of the sound intensity. This is the reason that heart sounds are weakly transmitted to the air at the chest wall. A stethoscope, which serves as an impedance matching device, is required to hear these sounds.

The examples which follow illustrate how these equations can be used to compute reflection and transmission amplitudes.

Example 10.1

Calculate the ratios of the pressure amplitudes and intensities of the reflected and transmitted sound waves from air to water relative to A_0 and I_0.

Using the values of Z from Table 10.1 along with Equations 10.6 and 10.7 we obtain for the pressure amplitude ratios:

$$\frac{R}{A_0} = \frac{1.48 \times 10^6 - 430}{1.48 \times 10^6 + 430} = 0.9994$$

$$\frac{T}{A_0} = \frac{2 \times (1.48 \times 10^6)}{1.48 \times 10^6 + 430} = 1.9994$$

We may also obtain (from Eq. 10.3) the ratios of the reflected and transmitted intensities:

$$\frac{IR}{I_0} = \frac{R^2}{2Z_1} \Big/ \frac{A^2_0}{2Z_1} = \left(\frac{R}{A_0}\right)^2 = (0.9994)^2 = 0.9988$$

$$\frac{IT}{I_0} = \frac{T^2}{2Z_2} \Big/ \frac{A^2_0}{2Z_1} = \frac{Z_1}{Z_2}\left(\frac{T}{A_0}\right)^2 = \frac{430}{1.48 \times 10^6}(1.9994)^2 = 0.0012$$

Example 10.2

Calculate the ratios of pressure amplitudes and intensities for sound waves from water into muscle.

Using the values from Table 10.1 in Equations 10.6 and 10.7 we obtain for the pressure ratios:

$$\frac{R}{A_0} = \frac{(1.64 - 1.48) \times 10^6}{(1.64 + 1.48) \times 10^6} = 0.0513$$

$$\frac{T}{A_0} = \frac{2\,(1.64 \times 10^6)}{(1.64 + 1.48) \times 10^6} = 1.0513$$

The ratios for the intensities (from Eq. 10.3) are then:

$$\frac{IR}{I_0} = \left(\frac{R}{A_0}\right)^2 = (0.0513)^2 = 0.0026$$

$$\frac{IT}{I_0} = \frac{Z_1}{Z_2}\left(\frac{T}{A_0}\right)^2 = \frac{1.48 \times 10^6}{1.64 \times 10^6}(1.0513)^2 = 0.9974$$

10.2 The Body as a Drum (Percussion in Medicine)

Percussion (tapping) has been used since the beginnings of civilization for various purposes such as testing whether a wall is solid or is a covering for a hiding place, and whether wine barrels are empty or full. The first recorded use of percussion on the body for diagnosis occurred in the eighteenth century. In 1761, L. Auenbrugger published a short (95 page) book, *On Percussion of the Chest*, which was based on his clinical observations over seven years of the different sounds he produced by striking the chests of patients in various places. Auenbrugger was an accomplished musician and his father had been an innkeeper. He probably learned the technique of percussion by tapping wine barrels at his father's inn, and his musical ear probably helped him interpret the sounds.

In his book, Auenbrugger described how to strike the chest with the fingers (Fig. 10.3), and he stated, "the sound thus elicited from the healthy chest resembles the stifled sound of a drum covered with a thick woolen cloth or other envelope." He discussed the sounds heard from healthy subjects and the sounds heard from patients with various pathological conditions. He stated that with percussion he could diagnose cancer, the presence of abnormal cavities in an organ, and those diseases involving fluids in the chest region. He confirmed many of these diagnoses by autopsy.

Auenbrugger's discovery was largely ignored until 1808 when his work, originally published in Latin, was translated to French with commentary by J. N. Corvisart. Percussion grew in popularity along with auscultation. It is still generally part of physical diagnosis.

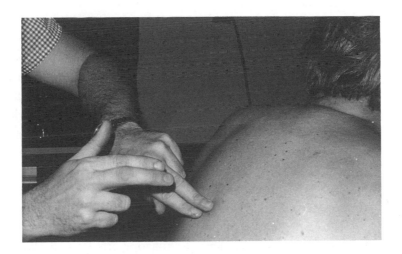

Figure 10.3. A method of inducing percussion of the chest from the back. The fingers of one hand are held against the skin and tapped with the fingers of the other hand.

10.3 The Stethoscope

Perhaps no symbol identifies the medical professional more readily than the ever present stethoscope. This simple "hearing aid" permits the doctor to listen to sounds originating inside the body, primarily in the heart and lungs but also in bone joints and partially obstructed arteries. Listening to these sounds with a stethoscope is called *mediate auscultation*, or usually just *auscultation*. Many sounds from the chest region can be useful in the diagnosis of disease. Prior to 1818, the only methods available for examination of the chest were feeling with the hand, percussion, and, occasionally, immediate auscultation with the ear directly on the chest. In *A Treatise on the Diseases of the Chest and on Mediate Auscultation* (1818), R. T. H. Laennec described the objections to putting the ear directly on the chest: "it is always inconvenient, both to the physician and the patient; in the case of females it is not only indelicate, but often impracticable; and for that class of persons found in the hospitals it is disgusting." (At that time, physicians routinely made house calls and examined and treated almost everyone of any means in his home. Only charity patients went to the hospital.)

Laennec used immediate auscultation until 1816 when he was examining a girl with general symptoms of a diseased heart. Because she was fat, young, and female, he felt that the usual examination methods were inappropriate. However, he recalled that if one end of a piece of wood is scratched with a pin, the sound can be heard well when the other end is held to the ear. He immediately rolled several pieces of paper into a cylinder and held one end to his ear and the other to the girl's chest above her heart. The results were dramatic and encouraged Laennec to improve this instrument. Eventually he developed a hollow wood cylinder 30 cm long with an inner diameter of approximately 1 cm and an outer diameter of about 4 cm. He called it a *stethoscope*, meaning "to view the chest." In his book he described his research on the stethoscope and his interpretation of the natural and pathological sounds of the lungs, heart, and voice.

The stethoscope currently in use is based on Laennec's original work. The main parts of a modern stethoscope are the bell, which is either open or closed by a thin diaphragm, the tubing, and the earpieces (Fig. 10.4).

The *open bell* serves to match the impedance between the skin and the air. It accumulates sounds from the contacted area. The patient's skin under the open bell behaves like a diaphragm. It has a natural resonant frequency at which it most effectively transmits heart sounds. The reso-

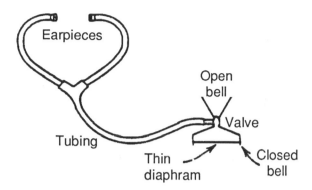

Figure 10.4. A schematic of the stethoscope.

nant frequency is controlled by the diameter of the bell and the pressure with which the bell is held on the skin. The tighter the skin is pulled, the higher its resonant frequency. The larger the bell diameter, the lower the skin's resonant frequency. It is possible to enhance the sound range of interest by changing the bell size and varying the pressure of the open bell against the skin, and thus the skin tension. A low-frequency heart murmur will not be heard if the stethoscope is pressed hard against the skin!

A *closed bell* is merely a bell with a diaphragm of known resonant frequency, usually high, that tunes out low-frequency sounds. Its resonant frequency is controlled by the same factors that control the frequency of the open bell pressed against the skin. The closed-bell stethoscope is primarily used for listening to lung sounds, which are of higher frequency than heart sounds. Fig. 10.5 shows the typical frequency ranges of heart and lung sounds.

What is the best shape for the bell? Since we are dealing with a system closed at the far end by a pressure-sensitive diaphragm—the eardrum—it is desirable to have a bell with as small a volume as possible. The smaller the volume of gas in the bell, the greater the pressure change for a given movement of the diaphragm at the end of the bell.

The volume of the tubes should also be small, and there should be little frictional loss of sound to the walls of the tubes. The small volume restriction suggests short, small-diameter tubes, while the low-friction restriction suggests large-diameter tubes. Thus if the diameter of the tube is too small, frictional losses dominate. If the diameter is too large, the moving air volume is too great. In both cases the efficiency is reduced. Below about 100 Hz the tube length does not greatly affect the efficiency,

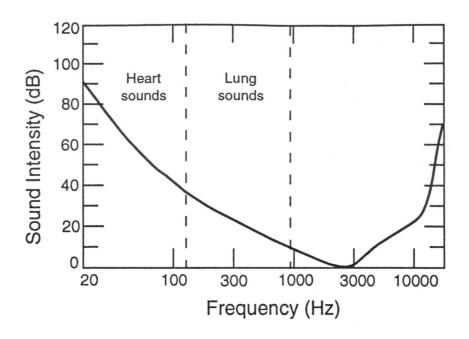

Figure 10.5. Most of the heart sounds are of low frequency in the region where the sensitivity of the ear is poor. Lung sounds generally have higher frequencies. The curve represents the threshold of hearing for a good ear. Some of the heart and lung sound levels are below this threshold.

but above this frequency the efficiency decreases as the tube is lengthened. At 200 Hz, 15 dB are lost in changing from a tube 7.5 cm long to a tube 66 cm long. A compromise is a tube with a length of about 25 cm and a diameter of 0.3 cm.

The earpieces should fit snugly in the ear because air leaks reduce the sounds heard. The lower the frequency, the more significant the leak. Leaks also allow background noise to enter the ear. The earpieces are usually designed to follow the slightly forward slant of the ear canals.

10.10

PROBLEM

What is the difference between percussion and auscultation?

What factors affect the selection of the diameter and length of the tube used for a stethoscope?

10.4 The Production of Speech (Phonation)

Normal speech sounds are produced by modulating an outward flow of air. (It is possible, though difficult, to make speech sounds while inspiring air. It is said that ventriloquists use this technique.) For most sounds the lungs furnish the stream of air, which flows through the *vocal folds* (cords), sometimes called the *glottis*, and causes them to vibrate, thereby modulating the air stream. The air then passes through several vocal cavities before it exits the body through the mouth and to a slight degree through the nostrils (Fig. 10.6). Speech sounds produced in this way are called *voiced sounds*. Sounds produced in the oral portion of the vocal tract without the use of the vocal folds are called *unvoiced sounds*. Examples are *p, t, k, s, th,* and *ch*. The *p, t,* and *k* are known as *plosive* sounds while the *s, f,* and *th* are known as *fricative* sounds; *ch* is a combination of the two types. The unvoiced sounds involve air flow through constrictions or past edges formed by the tongue, teeth, lips, and palate. Try making some of these sounds and notice how you use your tongue, teeth, and lips in the process.

In this section we consider only the production of voiced sounds. Since the vocal mechanism is far too complex to examine in detail from the acoustic point of view, we use a model of the vocal tract (Fig. 10.7). In this model, the sound is produced at the vocal folds and is selectively modified or filtered by three cavities. (This type of model is sometimes called a source-filter model.)

The vocal folds are located within the *larynx* (or Adam's apple) which is inside the *trachea*, or windpipe (Figs. 10.6 and 10.7). Fig. 10.8 shows the vocal folds as viewed from above, and Fig. 10.9 shows a vertical cross section of the larynx as seen from the front. During normal breathing the folds are widely separated, forming a large triangular opening (Fig. 10.8a). In the production of the vocal sounds the vocal

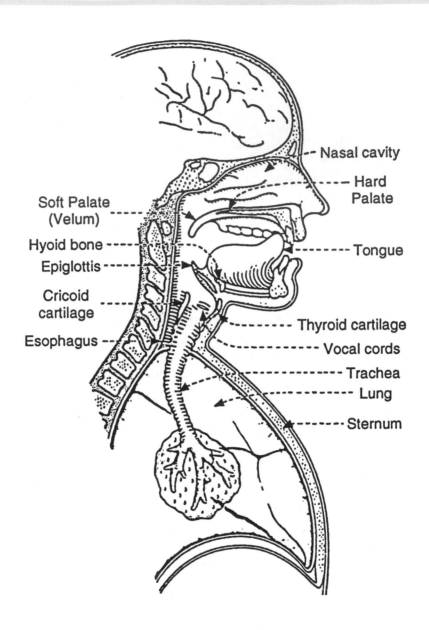

Figure 10.6. Schematic of human vocal mechanism. (From J. L. Flanagan, *Speech Analysis, Synthesis and Perception*, 2nd ed., Springer-Verlag, Heidelberg, 1972, p.10.)

folds are drawn close together by muscles (Fig. 10.8b), the air in the lungs is exhaled, the pressure below the vocal folds rises, and the closed folds are forced apart. The resulting rapid upward flow of air causes a decrease in pressure between the folds due to the *Bernoulli effect* (see Sections 7.3 and 8.6). The decrease in pressure, along with the elastic

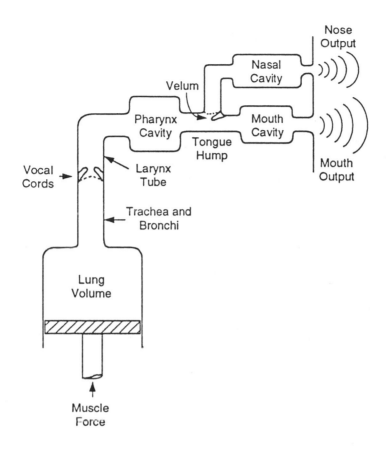

Figure 10.7. Model of human vocal mechanism. (From J. L. Flanagan, *Speech Analysis, Synthesis and Perception*, 2nd ed., Springer-Verlag, Heidelberg, 1972, p. 24.)

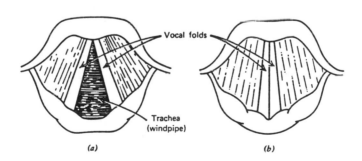

Figure 10.8. Sketch of the vocal folds as seen with a mirror held in the back of the throat. (a) The normal opening during aspiration. The dark area is the windpipe below the folds. (b) During phonation (speech). The vocal folds are drawn close together and vibrate as air is forced between them.

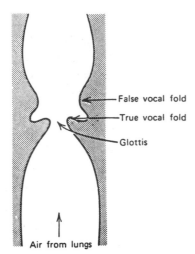

False vocal fold

True vocal fold

Glottis

Air from lungs

Figure 10.9. A cross-sectional view of the larynx as seen from the front.

forces in the tissues, causes the folds to move together, partially block-ing the passage and thus reducing the air velocity. This reduced air velocity increases the pressure below the folds and causes the process to begin again. The result is called a *glottal sound wave*.

The fundamental frequency of the resulting complex vibration depends on the mass and tension of the vocal folds. Men, who have longer and heavier vocal folds than women, have a typical fundamental frequency of about 125 Hz: the typical fundamental frequency for women is about one octave (a factor of two) higher, or 250 Hz. The lowest frequency that can be produced by a bass singer is about 64 Hz (corresponding to low C on the musical scale), and the highest frequency that a soprano can produce is about 2048 Hz (five octaves, or a factor of $2^5 = 32$, above low C).

The glottal sound wave passes through several vocal cavities—the pharyngeal (throat), oral, and nasal cavities that further change the sound of the wave that is emitted (see Fig. 10.7). The shape of the throat and nasal cavities are pretty well fixed for each individual and to a large extent determine the sound of the voice. They cannot be changed much voluntarily except if you hold your nose and talk. Swelling of tissues due to a head cold will alter them and cause a change in the voice. The oral cavity changes shape through the movement of the tongue, lower jaw, soft palate, and cheeks to determine voiced sounds. The tongue, palate, and cheeks in particular select the desired sounds out of the com-plicated periodic wave. If you think about the production of sound when you are speaking, you can feel your tongue change and your cheeks reshape when speaking the vowel sounds and some of the con-sonant sounds.

The sound produced in speaking can be greatly altered when a cleft palate exists. This is a birth defect in which there is an opening in the roof of the mouth; it may involve both the hard and the soft palates (Fig 10.6). Sometimes the upper lip is also split. Surgery or prosthetic devices can be used to remedy the condition.

Figure 10.10a shows schematically the air velocity in the glottal region, Figure 10.10b shows the vocal tract modification, and Fig. 10.10c shows the resultant radiative sound wave. It is equally possible to analyze the complex sound wave into frequency components and to determine the amplitudes of these components by a method called Fourier analysis (Fig. 10.10a'). Figure 10.10b' shows the sound transmission characteristics of the vocal tract. The action of the transmission characteristics of the vocal tract on the frequency components of the glottal sound produces the sound spectrum shown in Fig. 10.10c. Human speech is composed

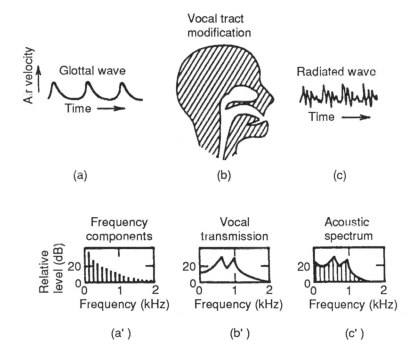

Figure 10.10. Two ways of viewing production of speech sounds. The glottal sound wave (a) is modified by the vocal tract (b) to produce a radiative wave (c). The amplitude of the frequency components of the glottal wave can be obtained (a'). They are modified by a function that represents the characteristics of the vocal tract (b') to produce the acoustic spectrum of the radiated wave (c'). (Adapted from Gunnar Fant, *Acoustic Theory of Speech Production*, © 1970 Mouton, The Hague/Paris, p. 19.)

from a rich variety of glottal sounds specifically modified by the vocal tract. Control is timed by the central nervous system.

10.4.1 Voice Word Power

When the sentence "Joe took Father's workbench out" is spoken in a normal voice, the kinetic and potential energy in the resultant sound is in the range 30 to 40 µJ. This is a very tiny amount of energy. The time needed to say the sentence is about 2 s, thus the average power is about 15 to 20 µW. A person could talk continuously for 100 years and still not produce the sound energy equivalent to the heat energy needed to bring a cup of water to the boil (the latter takes about 8×10^4 J—check it out!).

We can hear the spoken word even though the energy is small because of the great sensitivity of the ear. The vowel sounds contain much more power than the consonant sounds. Thus vowel sounds are easier to hear and understand than consonant sounds. In one study the sound intensity between the vowel sound in *aw*l and the consonant sound *th* in *th*in was found to be 680:1. This corresponds to $L_I = 28$ dB!

We normally think of the sound produced by belching as having no practical value. However, patients who have had their larynx removed can be taught to swallow air and to use controlled belching as an artificial larynx to produce voice sounds.

10.12

PROBLEM

What is the average power of the human voice?

10.13

PROBLEM

Why are vowel sounds easier to hear than consonant sounds?

10.14

PROBLEM

What is the average L_I (dB) and power level (μW) for normal speech?

10.15

PROBLEM

Show that the energy required to raise a cup of water (237 ml) from room temperature to the boiling point is larger than the energy expended through normal speech continued for 100 years. (1 ml = 10^{-6} m^3)

10.4.2 Looking at the Vocal Cords

Sometimes loss of voice, voice disorders, or hoarseness develops due to an obstruction or polyp on the vocal folds. This calls for examination of the vocal folds, which may be done in several ways. One method, called *indirect laryngoscopy*, uses a bright light source directed into the throat where an angled mirror reflects the light onto the folds. The light reflected back from the folds can be directed into the viewer's eyes. Sometimes this does not give enough view and *direct laryngoscopy* is used. This method entails a flexible endoscope inserted through the mouth into the throat region for a better examination. Other methods used to visualize the vocal folds include *magnetic resonance imaging (MRI)*, *ultrasound*, and various radiological procedures (e.g., x-ray, CT scans, etc.).

11

Physics of the Ear and Hearing

Speech and hearing are the most important means by which we communicate with others. Through hearing we receive speech sounds from others and also listen to ourselves! In some ways it is more of a handicap to be born completely deaf than to be born blind. Any child who cannot hear the sounds from his or her own vocal cords cannot learn to talk without special training. In earlier times, a child deaf from birth was also mute, or dumb, and since so much of our learning takes place through hearing, the person often was not educated. (This might be the origin of the use of the word "dumb" to indicate ignorant or stupid.) It was not until the sixteenth century that people first realized that the inability of a deaf child to talk was fundamentally related to the deafness. In the early nineteenth century, special schools were established for deaf-mutes. While deaf people can now be taught to talk, their voices usually sound abnormal since they have no easy way to compare the voice sounds they make to the voice sounds produced by other people.

If a sound is loud enough, it can be "heard" by deaf persons through the sense of touch; they may feel vibrations of exposed hairs on their body and thus "hear" the loud sound through the nerve sensors at the roots of the hairs. We discuss shortly how the ear uses a sophisticated refinement of this technique to hear sounds billions of times weaker.

Medical specialists concerned with the ear and hearing are: (1) the *otolaryngologist*, an M.D. who specializes in diseases of the ear and throat; (2) the *otorhinolaryngologist*, an M.D. who specializes in diseases of the ear, nose, and throat (also called an ENT specialist); and (3) the *audiologist*, a non-M.D. who specializes in measuring hearing response, diagnosing hearing disorders through hearing tests, and rehabilitating those with varying degrees of hearing loss.

The sense of hearing is in some ways more remarkable than the sense of vision. We can hear a range of sound intensities of over a million million (10^{12}), or 100 times greater than the range of light intensities the eye can handle (see Chapter 12, *Physics of the Eye and Vision*). The ear can hear frequencies that vary by a factor of 1000, while the frequencies of light that the eye can detect vary by only a factor of 2.

The sense of hearing involves: (1) the mechanical system that gathers and transmits the sound information so it can stimulate the hair cells in the cochlea; (2) the sensors that produce the action potentials in the auditory nerves; and (3) the auditory cortex, the part of the brain that decodes and interprets the signals from the auditory nerves. Deafness or hearing loss results if any of these parts malfunction. While they all involve physics, we know much more about the physics of the mechanical system than about the physics of the other parts. In this chapter we deal with the sense of hearing only up to the auditory nerve.

11.1 The Ear and Hearing

The ear is a cleverly designed converter of very weak mechanical sound waves in air into electrical pulses in the auditory nerve. Fig. 11.1 shows most of the structures of the ear that are involved with hearing. The ear is usually divided into three areas: the outer ear, the middle ear, and the inner ear. What we commonly call the ear (the appendage we use to help hold up our glasses) has no essential role in hearing.

The *outer ear* consists of the ear canal, which terminates at the eardrum (tympanic membrane). The *middle ear* includes the three small bones (ossicles) and an opening to the mouth (Eustachian tube). The *inner ear* consists of the fluid-filled, spiral-shaped cochlea containing the organ of Corti. Hair cells in the organ of Corti convert

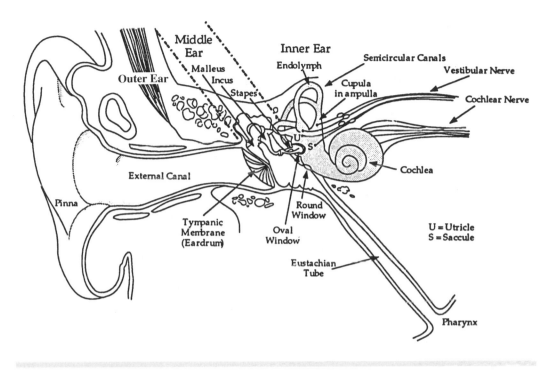

Figure 11.1. Cross section of the ear showing the outer, middle, and inner ear. The inner ear consists of the cochlea with the oval and round windows and the chambers inside the cochlea. Nerve connections carry auditory information to the brain where it is interpreted. The inner ear along with the additional attachment of the vestibular (balance) sensors is called the membraneous labyrinth.

vibrations of sound waves hitting the eardrum into nerve pulses that inform the auditory cortex of our brain of these sound waves. The inner ear is part of the labyrinth which also includes the sensors of the vestibular (sense of balance) system. The latter provides the brain with electrical signals that contain positional information of the head with respect to the direction of gravity and angular and linear motion information (see Section 11.10).

One of the first medical physicists to study the physics of the ear and hearing was Hermann von Helmholtz (1821–1894). He developed the first modern theory of how the ear works. His work was expanded and extended by Georg Von Bekesey (1900–1970), a communications engineer who became interested in the function of the ear as part of the communication system. Von Bekesey received a Nobel Prize in 1961 for his contributions to the understanding of the ear.

11.1

PROBLEM

Over what range of sound intensities can the normal ear hear?

11.2

PROBLEM

What are the three main components involved in the sense of hearing?

11.3

PROBLEM

What are the three areas of the ear?

11.2 The Outer Ear

The outer ear does not refer, as you might think, to the visible part of the ear, which in medical jargon is called the *external auricle* or *pinna*. The outer ear is the *external auditory canal*, which terminates at the eardrum. The pinna is the least important part of the hearing system; it aids only slightly in funneling sound waves into the canal and can be completely removed with no noticeable loss in hearing, although its removal will not help anyone's appearance!

Many animals, including humans, have muscles to move the ears. A horse has 17 muscles for each ear, while humans have 9 residual muscles. Most people do not have the nerve circuitry to activate these muscles. Those of us who have control over the movement of these muscles (who can wiggle our ears) find that its primary use is to entertain children[1]. In some animals the pinna does play a role in collecting sound energy and concentrating it to the eardrum. The large ears of the elephant and of many desert animals also serve the important function of helping the animal cool itself.

The external auditory canal, in addition to being a storage place for earwax, serves to increase the ear's sensitivity in the region of 3000 to 4000 Hz. The canal is about 2.5 cm long (1 inch) and about the diameter of a pencil. You can think of the canal as an organ pipe closed at one end (length = $\lambda/4$) with a resonant frequency of about 3300 Hz (λ = 10 cm). Notice that the sensitivity of the ear is best in this region (Fig. 11.2).

The eardrum, or tympanic membrane, is about 0.1 mm thick (paper-thin) and has an area of about 65 mm^2. It couples the vibrations in the air to the small bones in the middle ear. Because of the off-center attachment of the malleus (Fig. 11.3a), the eardrum does not vibrate

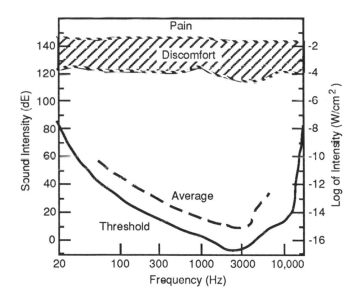

Figure 11.2. The sensitivity of the ear. The solid curve is the threshold of hearing for a young person with good hearing. Zero decibels occurs at 1000 Hz. The "average" curve is the average threshold for all people, young and old. Both axes—horizontal and vertical—are logarithmic scales.

---

[1]One of the authors (JRC) can wiggle each ear independently. Some people think this is his most outstanding accomplishment.

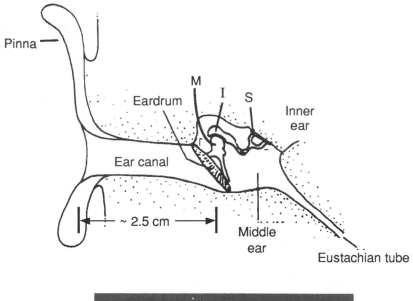

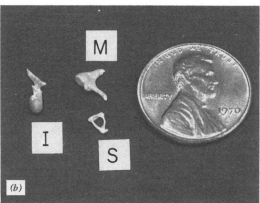

Figure 11.3. The ossicles of the middle ear. (a) A schematic cross section showing the stapes S, the malleus M, and the incus I. (b) Photographed next to a penny. Note the small size of the ossicles.

symmetrically like a drumhead. The motion of the eardrum was studied extensively by Von Bekesey who used cadaver ears. Recent studies using a sophisticated nuclear physics technique (the Mossbauer effect) to study the motion of the eardrum in the living ear indicate that many of the values Von Bekesey obtained were in error because of changes in tissue elasticity after death. However, it is clear that the actual movement of the eardrum is exceedingly small since it must be less than the movement of the air molecules in the sound wave. This movement at the threshold of hearing at 3000 Hz (see Figure 11.2) is about 10^{-11} m, less than the diameter of a hydrogen atom! At the threshold of hearing, at the lowest frequencies that we can hear (~20 Hz), the motion of the eardrum may be as large as 10^{-7} cm. This is still less than the wavelength of visible light. (See Problems 10.3 and 10.4.)

It is possible for sound pressures above 160 dB to rupture the eardrum. A ruptured eardrum, being made of living cells, normally heals similarly to other living tissue.

11.4

PROBLEM

From Fig. 11.2 find:
(a) the frequency at which the ear is most sensitive.
(b) the minimum intensity (in watts per square centimeter) that a person can hear at 100 Hz.

11.5

PROBLEM

A static pressure of 8×10^3 Pa across the eardrum can cause it to rupture. How does this compare to the sound pressure from a 160-dB sound that can also cause the eardrum to rupture?

11.3 The Middle Ear

The dominant features of the middle ear are the three small bones (ossicles), which are shown in Fig. 11.1 and in more detail in Fig. 11.3. These bones are full adult size before birth. (The fetus can hear while it is still in the womb and there are indications that it learns the sound of its mother's voice before being born.) The ossicles play an important role in matching the impedance at the eardrum to the impedance of the liquid-filled chambers of the inner ear. The ossicles are named after the objects they resemble: the *malleus* (hammer), the *incus* (anvil), and the *stapes* (stirrup). They are arranged so that they efficiently transmit vibrations from the eardrum to the inner ear. They transmit poorly vibrations in the skull, even the large vibrations from the vocal cords.[2] You hear your own voice primarily by transmission of sound through the air. Try plugging both your ears and listen to the reduction in your sound volume. (It is best to do this while you are alone, and preferably not in the library.)

The pressure of the sound on the eardrum is amplified as a result of the lever action of the ossicles. There is an even greater amplification due to the relatively large area of the eardrum compared to the area of the oval window. The lever amplification is shown schematically in Fig 11.4b. The model shows that the torque produced by the force F_m times the lever arm L_m is equal to the torque of the product of the force at the oval window F_o times its lever arm L_o.

$$F_m L_m = F_o L_o.$$

This expression can be further modified by writing the two forces in terms of pressures and respective areas at the eardrum and oval window, i.e., $F_m = P_m A_m$ and $F_o = P_o A_o$. Thus:

$$P_m A_m L_m = P_o A_o L_o.$$

and:

$$(P_o/P_m) = (A_m/A_o)(L_m/L_o).$$

The lever action amplifies the force by a factor of about $(L_m/L_o) = 1.3$. The ratio of the effective area of the eardrum to that of the base of the stapes is $(A_m/A_o) = 15$. This gain plus the lever gain of 1.3 results in a total gain in pressure of about 20! Converted to dB, the middle ear provides nearly 26 dB gain in the pressure ratio. This improves hearing efficiency and sensitivity. Another way of looking at the gain is by pointing out that the gain is primarily because the middle ear provides a good impedance match between the sound that travels in air as it enters the ear and the sound vibrations produced in the fluid of the internal ear. By impedance match we mean that the middle ear provides an efficient coupling of the sound energy in the air at the tympanic membrane into sound energy in the liquid of the cochlea. If the sound were to strike

---

[2]You can easily feel the vocal cord vibrations with your fingertips on your larynx, or Adam's apple. This phenomenon is used to help teach deaf children to speak.

the oval window directly, about 99.9% of it would be reflected, resulting in a 30 dB loss of intensity instead of the 26 dB gain provided by the middle ear.

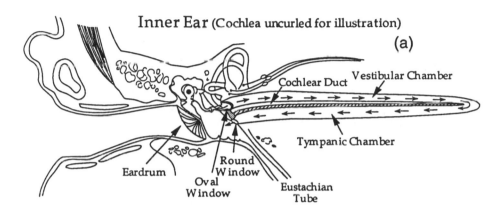

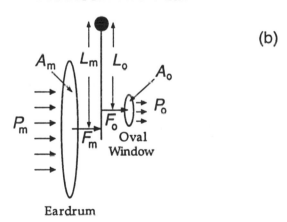

Figure 11.4. (a) Schematic view of the middle ear and the unrolled cochlea. The sound vibrations at the tympanic membrane rotate the malleus and incus about a common axis perpendicular to the page and marked with a dot. The stapes footplate moves the oval window membrane in and out with resulting transmission of the pressure changes into the cochlea. The pressure at the oval window is significantly larger (about a factor of 20) than that at the eardrum. The resulting vibration inside the cochlea stimulates small hair cells that provide the electrical nerve signals. (b) The lower diagram illustrates the impedance matching action of the middle ear, which takes place because of the difference in area of the eardrum and the oval window, and the lever ratio (L_m/L_o).

In the middle ear, the factors that affect the impedance are primarily the stiffness of the eardrum and its mass. The impedance match of the ear is fairly good from about 400 to 4000 Hz; below 400 Hz the "spring" is too stiff and above 4000 Hz the mass of the ear drum is too great. The middle ear aids the impedance match by amplifying the pressure by the lever and piston action described above. We discuss measurements of the middle ear in Section 11.7.

Loud sounds can damage the sense of hearing. The body has a mechanism to provide some protection against loud sounds. The stapedius muscle attached to the stapes and the tensor tympani muscle attached to the handle of the malleus play an important role in protecting the ear against loud sounds. A loud sound causes the muscles in the middle ear to pull sideways on the ossicles and reduce the sound intensity reaching the inner ear. A decrease of 15 dB is possible by this means. However, it takes about 15 ms or longer for these muscles to react, and damage may be done in this brief period. Persons living or working in an environment of loud sounds permanently lose some of their hearing sensitivity if they do not wear ear protectors. Noise pollution is not only unpleasant; it can result in permanent physiological damage to the hearing mechanism. The sound levels of some common sounds are given in Table 10.2.

11.6

PROBLEM

(a) Over what frequency range is the impedance of the ear fairly well matched to the acoustic impedance of the air?

11.4 The Eustachian Tubes

The Eustachian tubes connect each middle ear to the back of your mouth. They serve as a drainage path for fluids generated in the middle ear. As they open momentarily, they allow the pressure in the middle ear to equalize with the atmosphere. They should remain closed most of the time. If they decide to remain completely closed or open for many hours, they can produce physiological problems. The movement of the muscles in the face during swallowing, yawning, or chewing will usually cause a momentary opening of the Eustachian tube. During this momentary opening the pressure in the middle ear equalizes to that of atmospheric pressure. Sometimes you hear a popping sound in one or both ears as the eardrums return to their normal position.

The pressure equalization may occur spontaneously without jaw movement if the surrounding air pressure is decreasing, such as when you are going up in the elevator of a tall building or after taking off in an airplane. The Eustachian tube is smaller in diameter than indicated in Figures 11.1 and 11.4. Normally it is closed rather than open as shown. Air in the middle ear is normally slowly absorbed into the tissues, lowering the pressure on the inner side of the eardrum. If for some reason the Eustachian tube does not open, the pressure difference deflects the eardrum inward and decreases the sensitivity of the ear. At about 8 kPa or 1/12 atmosphere across the eardrum, the pressure difference causes pain. Common reasons for the failure of this equalizing system are blockage of the Eustachian tube by viscous fluids from a head cold and the swelling of tissues around the entrance to the tube.

11.5 The Inner Ear

The inner ear, hidden deep within the hardest bone in the body, the petrous bone of the skull, is man's best-protected sense organ (Figures 11.1 and 11.4a). The inner ear consists of a small spiral-shaped, fluid-filled structure called the *cochlea*. When the eardrum moves, the ossicles of the middle ear cause the stapes to push on the flexible membrane covering the oval window of the cochlea and transmit pressure variations to the fluid in the cochlea (Fig. 11.4a). These vibrations cause motion in a flexible basilar membrane, which stimulates hair cells in the *organ of Corti*. The stimulated hair cells produce electrical pulses (action potentials). These signals go to the brain via the auditory nerve—a bundle of about 28,000 conductors that inform the brain which parts of the cochlea are stimulated by the sound (Figs. 11.1 and 11.5). The cochlear nerve provides information on both the frequency and the intensity of the sounds that we hear. Many aspects of the inner ear are still being studied.

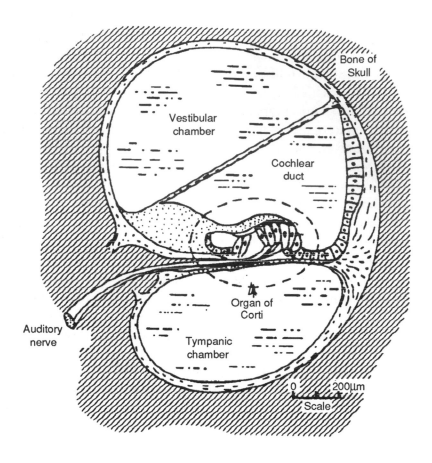

Figure 11.5. The chambers of the cochlea in cross section. The organ of Corti inside the dashed line is shown in more detail in Figure 11.6.

The cochlea is about the size of the tip of the little finger. If its spiral were straightened out, the cochlea would be about 3.4 cm long (Fig. 11.4a). It is divided into three small, fluid-filled chambers that run its full length as shown in cross-sectional view in Fig. 11.5. The oval window is on the end of the *vestibular chamber*, the middle chamber is the *cochlear duct*, and the third chamber is the *tympanic chamber*. The vestibular and tympanic chambers are interconnected at the tip of the spiral. The perilymph fluid in the two chambers is derived from cerebrospinal fluid (CSF) around the brain. Pressure produced at the oval window of the vestibular chamber by the stapes is transmitted via the vestibular chamber to the end of the spiral and then returns via the tympanic chamber. Since fluid is almost incompressible, the cochlea needs a "relief valve"; the flexible round window at the end of the tympanic chamber serves this purpose.

A sinusoidal sound wave of a particular frequency at the oval window produces a wave-like ripple in the basilar membrane of the cochlear duct (Fig. 11.6). The movement of the basilar membrane is a maximum at a particular point along its length. That point identifies the frequency of the incoming sound. Hair cells on the basilar membrane with the greatest movement send signals to the brain, which the brain interprets to be the stimulating frequency or pitch.

The stiffness of the basilar membrane changes by a factor of 10,000 from the oval window to the tip. It is most rigid near the oval window and has its greatest movement there for high frequencies. The low frequencies produce greatest movement at the tip. Hair cell sensors detect the motion of the basilar membrane. Thus, any auditory signal sent to the brain from a particular hair cell automatically specifies its frequency or pitch. Hair cells are discussed in more detail in Section 11.6.

The frequency spectrum is distributed logarithmically along the basilar membrane. That is, a distance of 3.5 to 4 mm corresponds to one octave or a factor of two change in the detected frequency. This distance is the same for low notes or high notes. Frequencies from 10,000 to 20,000 Hz use the same length of the basilar membrane as from 40 Hz to 80 Hz.

Studies of the minute magnetic fields from the brain (magnetoencephalography or MEG) have been used to map the location of magnetic

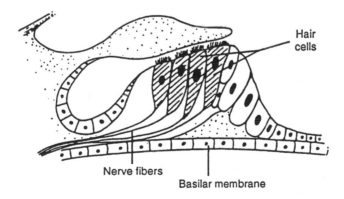

Figure 11.6. Simplified schematic of the organ of Corti showing the hair cells.

signals in the brain when different frequency sounds are heard. They are found to be distributed in a similar logarithmic manner. The logarithmic distribution on the basilar membrane is illustrated in Fig. 11.7 where 10 octaves of an unusual miniature piano keyboard are displayed along a magnified basilar membrane. This unusual piano can play notes from roughly 20 Hz to 20,000 Hz. The first eight octaves from 20 Hz to about 5,000 Hz occupy about two-thirds of the length of the basilar membrane. The remaining frequency spectrum from 5000 to 20,000 is squeezed into the remaining one-third of the thick end of the basilar membrane.

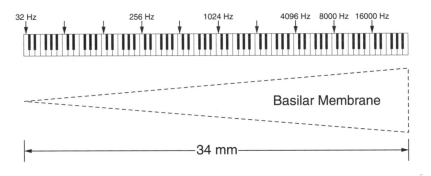

Figure 11.7 One can imagine the basilar membrane as a miniature piano with additional keys on each end to cover the large frequency range of 32 to 20,000 Hz. Each octave is represented by about 3–4 mm along the length of the basilar membrane. It would take thousands of additional "keys" to represent the true frequency discrimination of the ear. The keys on this imaginary piano are "played" by vibrations in the basilar membrane produced by the incoming sound wave.

The motions of the basilar membrane are about 10 times smaller in amplitude than the motions of the eardrum, which already are extremely small. Stimulation of nerves in the cochlear duct near the oval window indicates high-frequency sounds. Low-frequency sounds cause "large" motions and stimulate the nerves near the tip of the basilar membrane.

The transducers that convert the mechanical vibrations into electrical signals for the brain are located in the bases of the hair cells in the organ of Corti (Fig. 11.6). When a sound of 10,000 Hz is heard, the nerves located in the portion of the organ of Corti send a series of pulses that indicate which portion of the audible spectrum is being received.

11.6 Hair Cells Play the Dominant Role in the Detection of Sound

Hair cells of the organ of Corti are the primary converters of sound energy to an electrical nerve signal. Hair cells have a little tuft of "hair" made up of about 100 closely packed "hairs" a few microns in length. A schematic enlarged view of a cross section of a hair cell is shown in Fig. 11.8. The hairs are exposed to the fluid in the organ of Corti. As the basilar membrane moves under the influence of the sound pressure wave from the oval window, the hair cells move relative to the fluid. Friction with the fluid produces microscopic forces on the hairs, tending to bend them. If they are bent in an appropriate direction, they produce a voltage that can initiate an action potential.

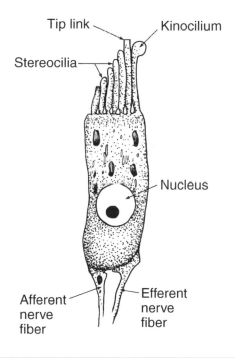

Figure 11.8 A sketch of the sound detection mechanism of the ear—the hair cell. Each human ear has over a million moving parts. These are the components of the hair cells—compact bundles of about 100 microscopic hairs a few microns long. A very slight motion of a hair cell in the appropriate direction produces an action potential that tells the auditory cortex of the brain that a particular frequency has been detected. [Illustration redrawn. Reprinted with permission from A. J. Hudspeth and V. S. Markin, "The Ear's Gears: Mechanico-electrical Transduction by Hair Cells," *Physics Today* 47(2): p. 23 (1994). Copyright 1994 American Institute of Physics.]

There are a variety of hair cells in the labyrinth. The sacculus and utriculus have about 15,000 and 30,000 hair cells, respectively, to measure linear accelerations. The three semicircular canals to measure angular acceleration have about 7,000 each. The cochlea has a row of inner hair cells and three rows of outer hair cells. Each row has about 4,000 hair cells. The inner row of hair cells running the full length (~3.4 cm) of the basilar membrane detect motions in the basilar membrane produced by the sound pressure wave from the oval window. They are the primary transducers of the sound wave into electrical signals.

An extremely small motion of a fraction of a nanometer (10^{-9} m) can produce a signal. Our ears cover a dynamic range of over a billion and make time measurements of a few microseconds. The roles of the outer hair cells are not well understood. They apparently play a role in mechanically amplifying weak incoming sound waves.

To explain the response of hair cells, it is apparent that some responses occur in times on the order of a few microseconds. Since action potentials and muscle responses are of the order of milliseconds, a much faster mechanism is needed. It appears that electrical contractions of molecules in a single hair cell are needed to account for such a fast response.

Hudspeth summarizes the cochlea as "...an evolutionary triumph of miniaturization...the most complex mechanical apparatus in the human body, with over a million essential moving parts...an acoustical amplifier and frequency analyzer compacted into the volume of a child's marble."

For a more detailed discussion of the physics of the ear see Chapters 2 and 3 of the book by Roederer in the reading list for this chapter.

11.7 Sensitivity of the Ears

The ear is not uniformly sensitive over the entire hearing range. Its best sensitivity is in the region of 2 to 5 kHz. The lower curve in Fig.11.2 (threshold) shows the average values for a young person with good hearing. The *average* line shows the levels at which half the people tested will detect the sounds. Notice that even a good ear needs about 30 dB more intensity to detect a sound at 100 Hz than to detect one at 1000 Hz.

Sensitivity changes with age. The highest frequency you can hear will decrease as you get older, and the level of sounds must be greater for you to hear them. There is a loss of hearing sensitivity associated with aging (called *presbycusis*). A person 45 years old typically cannot hear frequencies above 10 kHz and needs about 10 dB more intensity than was needed at age 20 to be able to hear a 4000 Hz note. A 25 dB loss in sensitivity in the frequencies above 2000 Hz usually has occurred by age 65. This loss is not serious for most activities. Hearing deteriorates more rapidly if the ears are subjected to continuous loud sounds. Some young people who play in rock bands have had serious hearing losses. Factory workers who work under very noisy conditions, without ear protection, have also shown measurable losses in hearing.

The property of sound we call *loudness* is a mental response to the physical property called *intensity*. The loudness of a sound is roughly proportional to the logarithm of its intensity and this effectively compresses the huge range of sound intensities to which the ear responds (12 orders of magnitude). In addition, the loudness of a sound depends strongly on its frequency. A sound of 30 Hz that is barely audible has the same perceived loudness as a barely audible sound of 4000 Hz, even though their sound intensity levels differ by a factor of about 1,000,000, or 60 dB. A special unit has been designed for loudness—the *phon*. One phon is the loudness of a one dB, 1000 Hz sound; 10 phons is the loudness of a 10 dB, 1000 Hz sound; and so forth. The loudness of a sound at another frequency is obtained by adjusting the intensity until it appears as loud as the known intensity 1000 Hz sound. Figure 11.9 shows typical curves of equal loudness at the threshold of hearing and at 40 and 60 phons. These curves vary for different individuals, but for people with normal hearing the curves have one characteristic in common: as the loudness increases, the curves become flatter. The threshold of feeling is about 100 dB at all frequencies.

The frequencies of most importance to us are those of the human voice. The shaded area in Fig. 11.9 indicates the general range of frequencies and sound levels of ordinary conversational speech. You can see that the ear is not optimized for the speech frequencies. However, it is possible to have a hearing loss of 40 dB and still hear most conversation. (In Section 11.9 we discuss deafness and hearing aids.)

If the ear were as sensitive at low frequencies as it is in the 3000 Hz region, we would be aware of many physiological noises, such as blood flow in the arteries in the head, movement of the joints, and possibly the small variations of pressure on the eardrum due to random motion of air molecules (Brownian motion). If you go into a special soundproof room used for testing hearing, you will be impressed by how many internal body sounds you hear. Most of these sounds are transmitted through bone conduction to the inner ear. These sounds are

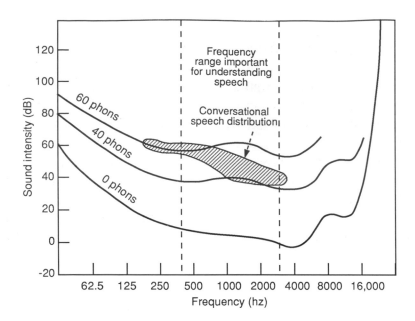

Figure 11.9. Curves of equal loudness at the threshold of hearing and at 40 and 60 phons. (Adapted from *Hearing and Deafness*, Third Edition, by Hallowell Davis and S. Richard Silverman, p. 28. Copyright 1947, 1960, 1970 by Holt, Rinehart and Winston, Inc. By permission of Holt, Rinehart and Winston, Inc.)

poorly detected by the ear, which is optimized for sounds coming via the eardrum. In general, a sound must be about 40 dB more intense to be heard by bone conduction than to be heard by air conduction.

11.8 Testing Your Hearing

If you have a hearing problem and consult an "ear doctor"—an otologist or otolaryngologist—you may be sent to an audiologist to have your hearing tested. If you have a hearing loss, the audiologist will be able to determine whether it is curable; if it is not, your ability to use a hearing aid will be assessed.

There are three somewhat standard tests to assess one's hearing: (1) pure tone audiometry, (2) speech audiometry, and (3) impedance (immitance) measurements that evaluate the function of the middle ear. The first and third methods are described below. The second is concerned with using speech in a prescribed form to test the hearing; it is somewhat similar to pure tone audiometry and will not be discussed further.

11.8.1 Pure Tone Audiometry

The pure tone audiometry tests normally are done in a specially constructed soundproof testing room. Each ear is tested separately; test sounds can be sent to either ear through a comfortable headset. The subject is asked to give a sign when the test sound is heard. Selected frequencies from 250 to 8000 Hz are used. At each frequency the operator raises and lowers the volume until a consistent hearing threshold is obtained.

The hearing thresholds are then plotted on a chart and can be compared to normal hearing thresholds (Fig. 11.10a). The normal hearing

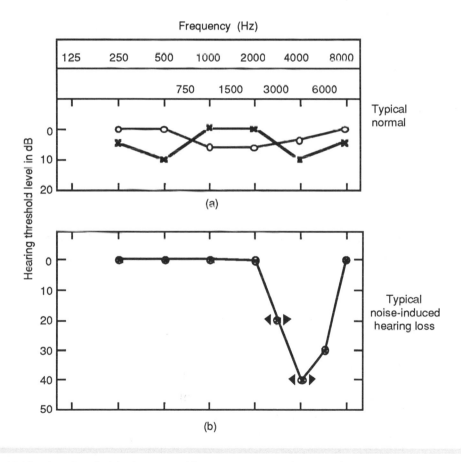

Figure 11.10. Hearing thresholds determined by a pure tone audiometric hearing test and plotted on a standard hearing chart. The open circles represent the threshold for air conduction in the right ear; the crosses are for the left ear. (a) A typical response of a person with normal hearing. (b) A typical noise-induced hearing loss in the region of 4000 Hz. The black triangles indicate thresholds for bone conduction.

threshold at each frequency is taken to be 0 dB. The chart may show a general loss in one or both ears. Usually a hearing loss is not uniform over all frequencies. Figure 11.10b shows the hearing threshold of a person with imperfect hearing. Notice the sharp hearing loss in both ears at about 4 kHz. In this case, the loss was due to partial nerve damage of the cochlea, an example of sensor-neural hearing loss (a condition where the air and bone hearing are nearly the same).

11.8. 2 Measuring Immitance of the Middle Ear

Acoustic impedance is the quantity that describes opposition to the flow of sound energy. We gave an example of this in Section 10.1.2. There we saw that for a sound wave in air striking water, only 0.1% of its incident intensity was transmitted into the water (a loss of 30 dB) while the reflected portion was 99.9%. The ear, because of its good impedance matching, results in about a 26 dB gain in pressure at the oval window. If the middle ear malfunctions because the ossicular chain is broken, the eardrum is perforated, or the Eustachian tube closes and the middle ear fills with fluid, the impedance of the middle ear changes.

If we know how much sound intensity reaches the eardrum, and if we can measure the amount reflected, then we can determine the amount transmitted since the sum of the reflected and transmitted intensities must equal the incident intensity. If the reflected intensity increases, then the transmitted portion decreases and vice versa. When the transmitted portion increases, this means that the impedance match is better.

The inverse of impedance is *admittance*. The word *immitance* includes both impedance or admittance. The impedance is considered to be mainly due to the stiffness of the eardrum. The inverse of stiffness is *compliance*, which is related to admittance.

Measurements of immitance of the middle ear are done with an instrument similar to that shown schematically in Fig. 11.11. The plug is inserted and sealed into the ear canal and sound energy from a small speaker driven at 220 Hz by an oscillator is directed onto the eardrum. The sound intensity reflected from the eardrum depends strongly on the stiffness of the eardrum. The stiffness depends on the air pressure in the ear canal. This is adjusted by the air pump and measured by the manometer. The pressure can be made either positive or negative. For the normal ear, reflected sound energy is least with equal pressure on both sides of the eardrum. At positive pressure, the eardrum is pushed

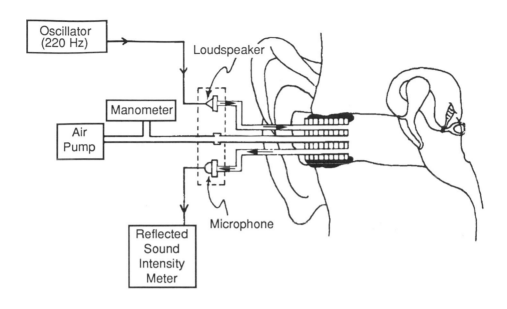

Figure 11.11. Schematic diagram of an instrument used to measure the immitance of the middle ear.

in toward the middle ear; with negative pressure, the eardrum deflects out from the middle ear (Fig. 11.12a). In either case more energy is reflected from the eardrum than for zero pressure. Measurements are made for both positive and negative pressures (upper scale of Fig. 11.12b). Often the inverse of the stiffness, the compliance, is measured versus the pressure in the ear canal (lower scale of Fig. 11.12b).

A plot of these data for a normal ear, called a *tympanogram*, is shown in Fig. 11.13a. It is Figure 11.12b rotated 90°. The stiffness is greatest at the extremes in pressure and much of the incident energy is reflected. The largest compliance occurs with zero pressure, the stiffness is least, and more energy is transmitted into the cochlea.

Several pathologies cause abnormal tympanograms. Figure 11.13b shows the middle ear, which is very stiff so the compliance is low even at zero pressure. Fig. 11.13c shows the high compliance of a flaccid eardrum, as might occur for an ossicular chain discontinuity.

Static compliance and acoustic reflex assessment are also measured with the apparatus shown in Fig. 11.12. Static compliance represents the ability of the middle ear to comply with the driving sound intensity at the tympanic membrane. Simply stated, it provides a measure of how much the eardrum stretches under positive pressure. Low values indicate stiff ears, while higher than normal values suggest a discontinuity in the ossicular chain.

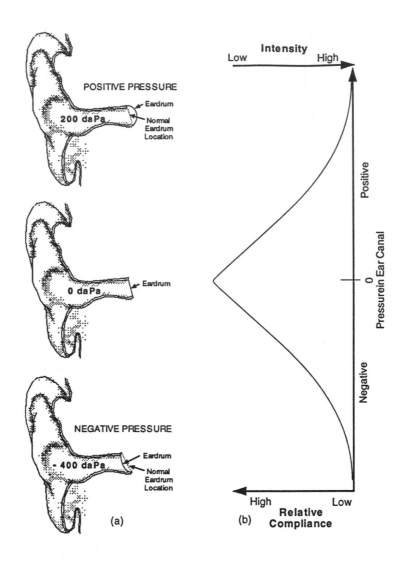

Figure 11.12. (a) Schematic view of the ear drum when the pressure in the outer ear canal is varied. (b) Schematic plot of the reflected intensity (top scale) or relative compliance (bottom scale) versus pressure (vertical scale) in the outer ear canal. (daPa = 10 Pa).

In the middle ear there are two muscles, the *tensor tympani* and the *stapedius*, which under reflex action protect the ear from loud sounds. The change in compliance is measured while the ear is stimulated by loud sounds. These acoustic reflex measurements are used to determine conductive disorders of hearing and, by inference, cochlear disorders.

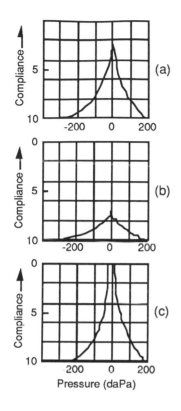

Figure 11.13. Several stylized tympanograms showing relative compliance versus pressure in the eardrum (daPa = 10 Pa). Note that one convention has the ordinate scale increasing opposite to the compliance. (a) Normal ear, (b) for a stiff ear drum, and (c) that observed for a flaccid ear drum as when the ossicular chain is interrupted. (Adapted from *Basic Audiometry*, T. J. Giolas, p. 66, © 1977 by Cliff Notes, Inc.)

11.9 Deafness and Hearing Aids

In 1985 it was estimated that 21 million persons in the United States were either deaf or hard of hearing. The frequency range most important for understanding conversational speech is from about 250 to 3000 Hz (Fig. 11.9). A person who is "deaf" above 4000 Hz but who has normal hearing in the speech frequencies is not considered deaf or even hard of hearing. However, that person should not spend a lot of money on good stereo equipment. Hearing handicaps are classified according to the average hearing threshold at 500, 1000, and 2000 Hz in the better ear. A person with a hearing threshold 30 dB above normal would probably not have a hearing problem. People with hearing thresholds of 90 dB are considered deaf or stone deaf. About 1% of the population have thresholds for speech frequencies greater than 55 dB and should use hearing aids. About 1.7% have a slight hearing handicap; they have problems with normal speech but have no difficulty with loud speech. Hearing problems increase with age.

The average sound level of speech is about 60 dB (cf. Table 10.2). We adjust the sound level of our speech unconsciously according to the noise level of our surroundings. Speech sound levels in a quiet room may be as low as 45 dB; at a noisy party they may be 90 dB. A person with a hearing loss of 45 dB in the 500 to 2000 Hz range may do all right (hearing-wise) at a cocktail party but hear very little of speech in a quiet room.

11.9.1 Conduction and Nerve Associated Hearing Loss

There are two common causes of reduced hearing: *conduction hearing loss*, in which the sound vibrations do not reach the inner ear, and *nerve hearing loss*, in which the sound reaches the inner ear but no nerve signals are sent to the brain.

Conduction hearing loss may be temporary due to a plug of wax blocking the eardrum or from fluid in the middle ear (otitis media). It may also be due to a solidification of the small bones in the middle ear. This condition can sometimes be corrected by an operation, called a *stapedectomy*, in which the stapes, which pushes on the oval window, is replaced with a piece of plastic. If a conduction hearing loss is not curable, a hearing aid can be used to transmit the sound through the bones of the skull to the inner ear. A nerve hearing loss may affect only a narrow band of frequencies or it may affect all frequencies. Until the 1980s there was no known aid for nerve hearing loss; however, cochlear implants are now available. These implants have wires connected to the auditory nerve. In some cases they have been successful in restoring hearing but the brain has to learn the new type of signals.

11.7

PROBLEM

What are two common causes of reduced hearing?

11.9.2 Hearing Aids

A hearing aid is a device that amplifies the incoming sounds. In its basic form, it consists of a microphone to detect sound, an amplifier to increase its energy, and a loudspeaker to deliver the increased energy to the ear (Fig. 11.14). Hearing aids increase both sound signals and background noise. The latter makes improvement in hearing a difficult task.

The hearing threshold that requires a person to use a hearing aid is quite variable. Some people lip-read to help themselves understand speech. The simplest hearing aid, which is quite effective if your hearing loss is not large, is to cup your hand behind your ear. This reflects about 6 to 8 dB of additional sound into your ear canal. In addition, you usually gain another 10 dB when the speaker notices you and speaks more loudly.

The earliest artificial hearing aid was the ear trumpet. It was shaped like a funnel and concentrated the sound energy at the ear. Ear trumpets were never common, probably due to the human tendency to hide our handicaps.

Early electronic hearing aids, before the invention of transistors, were bulky and the batteries wore out rapidly. The receiver and amplifier were worn on the body and the loud speaker was located in the ear.

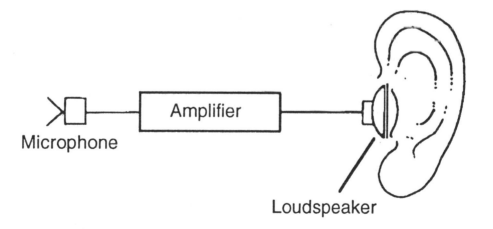

Figure 11.14. A hearing aid is really a small public address system with a microphone to pick up the sounds, an amplifier to increase the energy of the sound, and a loudspeaker to transmit the sound to the speaker's ear. Because the sound is directed into the auditory canal, very little power is needed.

A later form that enjoyed a reasonable popularity had the receiver and amplifier located in the frame of eyeglasses with a connection to the loudspeaker in the ear. A third type of hearing aid is located behind the ear; the most popular type is located in the ear. This reduction in size is largely due to miniaturization of the electronic circuits.

Electronic hearing aids are in common use today. Their most successful use is for hearing losses of 40 to 85 dB. For less than a 40 dB hearing loss, the occasional use of an aid may suffice. Hearing losses greater than 85 dB are helped little by hearing aids. Additional training with lip reading and even sign language may be necessary. For a person born deaf the problem of rehabilitation is more severe, for that person has never developed communication skills and more rigorous therapy must be used.

While it is possible to amplify the sound more than 90 dB, the pain threshold is the same as that for a person with normal hearing (about 100 to 120 dB). Thus, there is a practical upper limit on the sound output from an electronic hearing aid. This explains the limited usefulness of hearing aids for hearing losses of more than 80 to 85 dB. Hearing aids cannot return hearing to normal. They can only help compensate for the hearing loss. For example, an abrupt hearing loss above 3000 Hz cannot be completely corrected with a hearing aid. Most hearing aids have controls that permit the wearer to adjust the frequency response and the amplification, but the ranges of both are very limited. Improvements in hearing aids are still being made. Hearing losses are nonlinear; therefore, amplification in the hearing aid must be nonlinear to better produce the sounds heard by normal ears.

Digital technology, which processes sound digitally like CDs, has now joined the hearing aid industry offering an alternative to the more usual analog-based hearing aids. Digital hearing aids are now available for those users seeking the most advanced technology in a hearing aid. Because of the microminiaturization of electronic circuitry, the digital hearing aid can fit comfortably in the ear and yet have the improved features that the digital technology provides. They can be programmed for the individual's ears and are capable of filtering desired sounds and suppressing undesired sounds. Digital aids are "smart" instruments and represent the continuing desire to provide higher quality hearing aids. Digital hearing aids generally do not have patient adjustable frequency controls. The audiologist who fits the hearing aid programs the amplification profile to best match the needs of the ear in which it is worn. One of the authors (JRC) has a digital hearing aid. It does not help his hearing above about 4 kHz since he is "stone deaf" in that region. Its most amazing performance was to still function normally after a complete wash cycle in his shirt pocket.

11.10 The Vestibular Sense System—Our Hidden Sense of Balance

We close this chapter by considering the vestibular system (Fig 11.1), which maintains our body's coordination. We are generally unaware of this sensory system. We become uncomfortably aware of it at times when we have motion sickness. In some congenital deaf people this sensory system is missing; they do not experience motion sickness. They can still maintain their balance by using other senses, such as sight and pressure signals.

The vestibular system has five separate motion sensors. The physical principles of the hair cells in the motion sensors are identical to those of the hair cells in the organ of Corti for hearing. All vertebrates have similar hair cell sensors for hearing. The 134,000 hair cells in the motion sensors is much greater than the 32,000 hair cells in the inner ear. The most obvious motion detectors are sensors for angular acceleration located inside the three semicircular canals. Each of these canals, surrounded by dense bone, lies in a plane that is perpendicular to the other two. The DNA that programmed this sophisticated construction must be admired. During the growth of the skull in the fetus, the orientation of the fetus is continually changing. It is doubtful if the best modern engineers can suggest how to construct such a small complex structure containing many thousands of sensors inside a solid material.

Each of the semicircular canals is filled with endolymph that is separated from the cupula, the small bulb at the end of the canal by a septum. The motion of the fluid during acceleration pushes on the septum to cause motion in the cupula where the hair cells are located.

There are two motion sensors for linear acceleration perpendicular to each other located near the base of the semicircular canals. In Fig.11.1 the utricle (U) senses acceleration in the horizontal and the saccule (S) senses acceleration in the vertical plane. Having motion sensors on opposite sides of the body provides some redundancy.

The vestibular system is silent to our conscious mind. Signals of the position of our head and its movement are usually not noticeable to us.

We can upset the system by being in free fall—"weightless"—such as in a space ship. Nearly all astronauts suffer from motion sickness. What is surprising is that in a day or two the motion sickness disappears and it is impossible to induce it by intentional motions of the head.

A more common way to confuse the motion sensors is to consume alcohol, which apparently changes the density of the fluids in the semi-circular canals. This can give the impression that the room is in motion. One should avoid walking while the room is in motion.

Meniere's disease of the middle ear is moderately common. It can cause vertigo—dizziness and the sensation of falling. It often can lead to deafness in the affected ear. Bacterial and viral infections of the inner ear can also cause vertigo.

Our knowledge of the vestibular system is still sketchy, although much was learned in the last two decades of the 20th century. For more details on hair cells, the cochlea, and the vestibular system see the three articles by Hudspeth in the reference list for this chapter.

12

Physics of the Eyes and Vision

Most of our knowledge of the world around us comes to us through our eyes. The helplessness we feel when caught in the dark in unfamiliar surroundings is a good indication of our dependence on vision. The sense of vision consists of three major components: (1) the eyes that focus an image from the outside world on the light-sensitive retina (Fig. 12.1), (2) the system of millions of nerves that carries the information deep into the brain, and (3) the *visual cortex*—that part of the brain where "it is all put together." Blindness results if any one of the parts does not function. Physics is involved in all three parts, but the physics of the first part is understood far better than the physics of the other two parts.

In this chapter we discuss the physics of the eye. While the eye has some striking similarities to a camera, a better analogy exists between the eye and a closed circuit color TV system (Fig. 12.2). The lens of the TV camera is analogous to the cornea and lens of the eye; the "signal cable" is the optic nerve, and the "viewing monitor" is the visual cortex. When the light is bright we see things in "living color." In dim light the eye operates like a supersensitive black-and-white TV camera

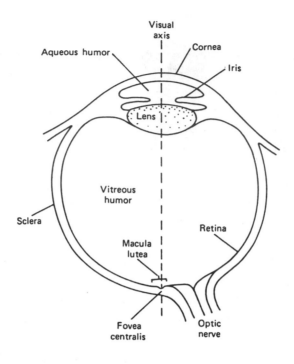

Figure 12.1. Cross section of the left eye as seen from above.

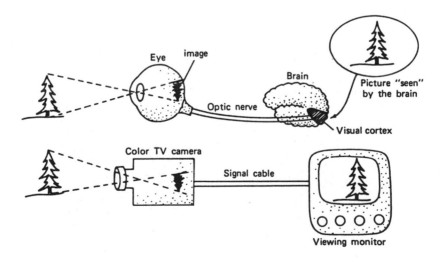

Figure 12.2. The sense of sight is in many ways similar to a closed circuit color TV system. It is superior in all respects except ease of placement.

and allows us to see objects with less than 0.1% of the light we need for color vision. This great difference in sensitivity is analogous to the difference between sensitive high-speed black-and-white film and the much less sensitive color film we use in our cameras.

Our optical system has the following special features, most of which are not available on even the most expensive cameras:

1. The eye can observe events over a very large angle while looking intently at an object directly ahead of it (Fig. 12.3).
2. Blinking provides the front lens (cornea) with a built-in lens cleaner and lubricator. (Each eyelid can be closed independently for communication with the opposite sex.)
3. A rapid automatic focusing system permits viewing objects as close as 20 cm (~8 in.) one second and distant objects the next. Under relaxed conditions the focus for normal eyes is set for "infinity" (distant viewing).
4. The eye can operate effectively over a range of light intensity of about 10 billion to one (10^{10}:1), brilliant daylight to very dark night.

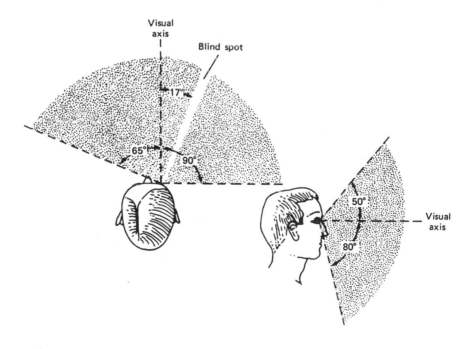

Figure 12.3. The eye looking straight ahead has a wide angle of vision.

5. The eye has an automatic aperture adjustment (the iris).

6. The cornea has a built-in scratch remover; even though it has no blood supply, it is made of living cells and can repair local damage.

7. The eye has a self-regulating pressure system that maintains its internal pressure at about 1.6 kPa (12 mm Hg) and thus keeps the eye in shape. If "dented," the eye rapidly returns to its original shape.

8. The eyes are mounted in a well-protected casing almost completely surrounded by bone, and each eye rests on a cushion of fat that reduces sharp shocks.

9. The image appears upside down on the light-sensitive retina at the back of the eyeball (Fig. 12.2), but the brain automatically corrects for this.

10. The brain blends the images from both eyes, giving us good depth perception and true three-dimensional viewing. If vision from one eye is lost, the vision from the remaining eye is adequate for most needs.

11. The muscles of the eye (Fig. 12.4) permit flexible movement up and down, sideways, and diagonally. After a little practice, the eyes can even be made to go in circles.

We discuss most of these features in more detail in this chapter.

Considering the sophistication of the eye mechanism, a surprisingly large percentage of people have good eyesight. These people are called *emmetropes*, but be careful when you use this term—some people may

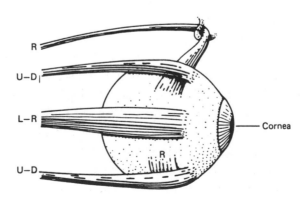

Figure 12.4. The six muscles of the right eye permit a wide variety of motion. The muscles work in pairs: one pair controls up and down movement (U-D), one pair controls left and right movement (L-R), and one pair controls rotation (R). The rotation muscles pass through bony loops. All six muscles are attached to the skull behind the eye.

misunderstand! The rest of us have noticeable vision imperfections and are called *ametropes*. If we consider an emmetrope to be anyone who needs correction of less than 0.5 D (diopters), 25% of young adults qualify; 65% fall within the ~1.0 D range. *Diopters* are a measure of the focusing power of the lens—it is numerically one meter divided by the focal length measured in meters.

Various medical specialists deal with problems of the eye. The most highly trained is the *ophthalmologist*, an M.D. who has taken three years of residency training in ophthalmology in addition to the M.D. training. The ophthalmologist is qualified to diagnose and treat any problem of the eye. Treatment may include surgery. The most common eye specialist, the *optometrist*, specializes in prescribing and fitting corrective lenses. An optometrist is now required to have six years of college, but is not licensed to treat diseases of the eye. An *ophthalmic technician* is a relatively new type of physician's assistant for ophthalmologists, and usually has had two or more years of college plus two years of on-the-job training. The technician can perform various eye tests and fit contact lenses (hard and soft), and also is trained in office management. An *optician* is a specialist in making lenses, fitting them in the frames, and fitting the frames to the patient. An optician usually has had two years of on-the-job training after high school. In some states opticians are required to be registered.

12.1 Focusing Elements of the Eye

The eye has two major focusing components: the *cornea*, which is the clear transparent bump on the front of the eye that does about two-thirds of the focusing, and the *lens*, which does the fine focusing. The cornea is a fixed focus element; the lens can change its shape and has the ability to focus objects at various distances.

The cornea focuses by bending (refracting) the light rays. The amount of bending depends on the curvatures of its surfaces and the speed of light in the lens compared with that in the surrounding material (relative index of refraction). The indexes of refraction of the cornea and other transparent parts of the eye are given in Table 12.1. When the cornea is underwater, it loses most of its focusing power because the index of refraction of the water (1.33) is close to that of the cornea (1.37). (Fish have a similar problem out of the water.) Divers keep air around the cornea by wearing a face mask. The index of refraction is

Table 12.1. The Indexes of Refraction of the Cornea and Other Optical Parts of the Eye

Part of the Eye	Index of Refraction
Cornea	1.34
Aqueous humor	1.33
Lens cover	1.38
Lens center	1.41
Vitreous humor	1.34

nearly constant for all corneas, but the curvature varies considerably from one person to another and is responsible for most of our defective vision. If the cornea is curved too much, the eye is near-sighted; not enough curvature results in farsightedness; and uneven curvature produces astigmatism. We discuss these defects more in Section 12.8. Nearly all of the focusing by the cornea is done at the front surface since the aqueous humor in contact with the back surface has nearly the same index of refraction as the cornea.

Since the living cells in the cornea are not supplied with oxygen by the blood, they must get their oxygen from the air. Having blood vessels in the cornea would not help our vision! The nutrients for the cells in the cornea are supplied by the aqueous humor that is in contact with its back surface. The aqueous humor contains all of the blood components except blood cells. We discuss this fluid in more detail in the next section.

If the cornea is scratched it will heal itself, but some other types of damage are more permanent. Some types of radiation (ultraviolet, neutrons, x-rays, etc.) can cause opacities to develop in the cornea that will block out light. It is possible to perform cornea transplants using corneas removed from donors shortly after death. Since the cells of the cornea have a low metabolism rate, rejection is not usually as much of a problem as in most organ transplants. Cornea transplants were the first organ transplants.

The lens has focusing properties at both its front surface and its back surface (Fig. 12.1). The lens is more curved in the back than in the front. It changes its focal strength by changing its curvature. The focusing power of the lens is considerably less than that of the cornea because it is surrounded by substances that have indexes of refraction close to its own. The effective index of refraction is thus only about 1.07. The lens is made up of layers somewhat like an onion, and all layers do not have the same index of refraction. The indexes of refraction in Table 12.1 are average values.

The lens has a flexible cover that is supported under tension by suspension fibers. When the focusing muscle of the eye is relaxed this tension keeps the lens somewhat flattened and adjusted to its lowest power, and the eye is focused on distant objects. The point at which distant objects are focused when the focusing muscle is relaxed is called the *far point*. For a near-sighted person the far point may be quite close to the eye. To focus on closer objects, the circular muscle around the lens contracts into a smaller circle and takes some or all of the tension off the lens. The lens oozes into a more spherical shape, primarily by becoming more curved in front. The lens then has a greater focusing power; objects that are closer to the eye are brought into focus at the retina. The closest point at which objects can be focused when the lens is its thickest is called the *near point*. Young children have very flexible lenses and can focus on very close objects. The ability to change the focal power of the eye is called *accommodation*. As people get older, their lenses lose some accommodation. *Presbyopia* (old sight) results when the lens has lost nearly all of its accommodation.

Not all animals focus by changing the shape of the lens; fish focus by moving the lens back and forth like we do with a camera. Perhaps they thus avoid presbyopia; you never see old fish wearing bifocals!

The lens, like the cornea, can be damaged by ultraviolet and other forms of radiation. It can develop *cataracts*, which destroy its clarity. It is possible to remove a damaged lens surgically. A plastic lens may be inserted to replace the damaged lens. If this is not possible for some reason, extra correction can be added to glasses to compensate for the loss of the lens. Of course no accommodation is then possible, and bifocals must be worn.

12.2 Some Other Elements of the Eye

The elements of the eye discussed in this section either are not directly involved with focusing the image or play a passive role in focusing. The retina plays such a dominant role in the function of the eye that it is discussed separately in Section 12.3. The components we discuss here are the iris and the pupil; the aqueous and vitreous humors; and the "housing" for the eye, the sclera.

The *pupil* is the opening in the center of the *iris* where light enters the lens. It appears black because essentially all of the light that enters is absorbed inside the eye. Under average light conditions, the opening is about 4 mm in diameter. It can change from about 3 mm in diameter in bright light to about 8 mm in diameter in dim light. The physiologic reason for this change in size is not clear. The maximum change of a factor of 7 in the opening area cannot begin to cover the huge range of light intensities the eye can handle, 10^{10}:1. The iris does not respond instantly to a change of light levels; about 300 s (5 min) are needed for it to fully open, and about 5 s are required for it to close as much as possible (Fig. 12.5).

It is believed that the iris aids the eye by increasing or decreasing incident light on the retina until the retina has adapted to the new lighting conditions. In addition, under bright light conditions it plays an important role in reducing lens defects. Camera bugs will realize that the small aperture increases the *depth of focus*, the range of distance over which objects are in satisfactory focus. You can demonstrate this effect by taking a piece of cardboard, making a hole about 1 mm in diameter in it, and looking through this hole at printed material under bright light. You can move the material very close to your eye and still be able to read it. The small aperture has increased your depth of focus. When you take away the pinhole, you will have to increase the distance to read the printing.

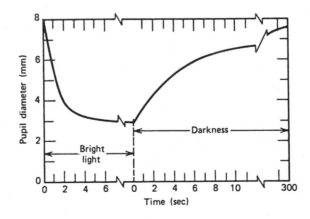

Figure 12.5. The pupil does not open and close rapidly. Note that the maximum opening is not attained until the eye has been in the dark 5 minutes.

The *aqueous humor* fills the space between the lens and the cornea. This fluid, mostly water, is continuously being produced, and the surplus escapes through a drain tube, the canal of Schlemm. Blockage of the drain tube results in increased pressure in the eye; this condition is called glaucoma. The aqueous humor contains many of the components of blood and provides nutrients to the nonvascularized cornea and lens. It maintains the internal pressure of the eye at about 1.6 kPa (12 mm Hg). (See Section 12.10 for a description of how this pressure is measured.) If you press on the eye, you find it is fairly stiff; you cannot indent it much. The reasons are that the fluids in the eye are incompressible at the pressure you use and that the covering of the eyeball does not stretch easily. When you rub your eyes, you greatly increase the internal pressure.

The *vitreous humor* is a clear jelly-like substance that fills the large space between the lens and the retina. It helps keep the shape of the eye fixed and is essentially permanent. It is sometimes called the vitreous body.

The *sclera* is the tough, white, light-tight covering over all of the eye except the cornea. The sclera is protected by a transparent coating called the *conjunctiva*.

12.3 The Retina—The Light Detector of The Eye

The retina, the light-sensitive part of the eye, converts the light images into electrical nerve impulses that are sent to the brain. While the role of the retina is similar to that of the film in a camera, a better analogy exists between the retina and the light-sensitive portion of a TV camera tube. Unlike film, the retina does not have to be replaced since a built-in system supplies the light-sensitive chemicals that in some way convert light into electrical nerve impulses. We do not understand completely the mechanisms involved, but we do know many of the characteristics of the photo receptors in the retina. In this section we discuss the physical aspects of the retina.

The absorption of a light photon in a photoreceptor triggers an electrical signal to the brain—an action potential. The energy of the photon is about 3 eV; the action potential has an energy millions of times greater. The light photon apparently causes a photochemical reaction in the photoreceptor which in some way initiates the action potential. The photon must be above a minimum energy to cause the reaction. Infrared photons have insufficient energy and thus are not seen. Ultraviolet photons have sufficient energy, but they are absorbed before they reach the retina and thus are not seen.

The retina covers the back half of the eyeball. While this large expanse permits useful "warning" vision over a large angle (Fig. 12.3), most vision is restricted to a small area called the *macula lutea*, or yellow spot. All detailed vision takes place in a very small area in the yellow spot (~0.3 mm in diameter) called the *fovea centralis* (Fig. 12.1).

The image on the retina is very small. An equation for determining the size of the image on the retina can be obtained by using the ratios of the lengths of the sides of similar triangles. In Fig. 12.6, O is the object size, I the image size, P the object distance, and Q the image distance, usually about 0.02 m (2 cm). Thus we can write O/P = I/Q or O/I = P/Q. Thus I = (Q/P)O.

12.1

PROBLEM

Assume that a fly is 3 mm (0.003 m) in diameter, and that the image distance for a normal eye can be taken to be Q = 0.02 m. Compute the size of the image formed on the retina of a fly sitting on a wall 3.0 m away. [Answer: I = 20 μm]

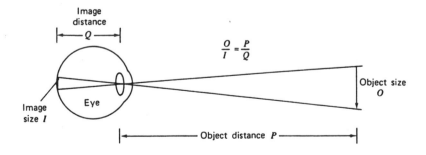

Figure 12.6. There is a simple relationship between the object and image sizes and the object and image distances. The image on the retina is small because of the short image distance of about 2 cm.

12.2

PROBLEM

If you are watching a football game from the end of a large stadium, what size will be the image of the football (L = 0.3 m) on your retina when the football is at the other end of the field, 150 m away from your seat? [Answer: 4×10^{-5} m]

There are two general types of photo receptors in the retina: the *cones* and the *rods*. Throughout most of the retina the cones and rods are not at the surface of the retina but lie behind several layers of nerve tissue through which the light must pass (Fig. 12.7). However, in the fovea centralis most of this nerve tissue is pushed to the side and there is a slight dip (*fovea* means pit). This decrease in nerve tissue aids our vision in this specialized area. The rods and cones are distributed symmetrically in all directions from the visual axis except in one region—the blind spot (Fig. 12.8).

The cones (about 6.5 million in each eye) are primarily used for daylight, or *photopic*, vision. With the cones, we can see fine detail and recognize different colors. The cones are primarily found in the fovea centralis, although some are scattered throughout the retina (Fig. 12.8).

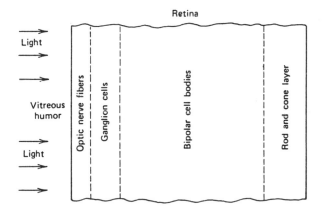

Figure 12.7. The light must pass through various cell layers to reach the rods and cones. In the fovea centralis much of the tissue is pushed to the side, permitting better detail vision in this area. Blood vessels also block the light to some rods and cones.

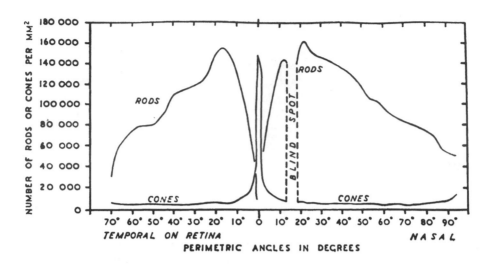

Figure 12.8. The distribution of the rods and cones in the retina of the left eye; notice the blind spot with no sensors. A perimetric angle of 30° indicates the location on the retina of the light source 30° to the left or right when the eye is looking straight ahead. (From M. H. Pirenne, *Vision and the Eye*, 2nd Ed., Chapman & Hall, Ltd., London, 1967, p. 32 as plotted from data of Osterberg, *Acta Opthal.*, Suppl. 6, 1935.)

We see later that the density of the cones in the fovea centralis determines the amount of detail we can resolve in an image. Each of the cones in the fovea has its own "telephone line" to the brain. In the rest of the retina several cones share one nerve fiber. The cones are not uniformly sensitive to all colors but have a maximum sensitivity at about 550 nm in the yellow-green region (Fig. 12.9). This corresponds quite well to the maximum in the solar spectrum at the earth's surface.

The rods are used for night, or *scotopic*, vision and for peripheral vision. They are much more abundant than the cones (about 120 million in each eye) and cover most of the retina. They are not uniformly distributed over the retina but have a maximum density at an angle of about 20° (Fig. 12.8). That is, if you are looking at the sky at night, the light from a faint star displaced 20° from your line of vision will fall on the most sensitive area of your retina. If you look directly toward the faint star, its image will fall on your fovea which has no rods and you will not see it.

Histological studies have indicated that hundreds of rods send their information to the same nerve fiber. This means that the ability to resolve two close sources of light in peripheral vision is poor. On the

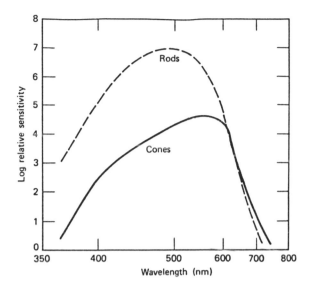

Figure 12.9. Rods are much more sensitive than cones. The vertical axis is a log scale; each division represents a factor of 10 in sensitivity. The best sensitivity of cones is at about 550 nm, while the best sensitivity of rods is at about 510 nm.

other hand, the great sensitivity of the rods and their great expanse permit us to recognize an object approaching from the side when we are looking straight ahead.

The rods are most sensitive to blue-green light ($\lambda = 510$ nm), a wavelength shorter than the optimum for the cones ($\lambda = 550$ nm). Figure 12.9 indicates that the rods and cones are equally sensitive to red light (650 to 700 nm). The curves in Fig. 12.9 are sometimes plotted so that both rods and cones have the same maximum sensitivity. This gives the erroneous impression that cone vision is better in red light than rod vision.

The eyes do not have their greatest sensitivity to light under photopic conditions. If the light level suddenly decreases by a factor of 1000, we are momentarily "in the dark"; but after a few minutes we are able to see many of the details that were not visible when it first became dark. This *dark adaptation* is apparently the time needed for the body to increase the supply of photosensitive chemicals to the rods and cones. The rate at which we dark adapt is shown in Fig. 12.10. The cones adapt most rapidly; after about 5 min the fovea centralis has reached its best sensitivity. The rods continue to dark adapt for 30 to 60 min, although most of their adaptation occurs in the first 15 min. It is possible to dark

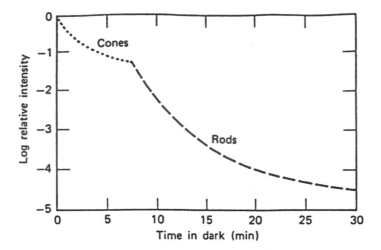

Figure 12.10. The eye dark adapts in two phases. The phase for the cones is completed in 5 to 10 min; the phase for the rods lasts for over 30 min. The increase in sensitivity is a factor of over 10,000.

adapt by wearing red goggles that limit the incident light to the red region of the spectrum. You can dark adapt one eye by closing it; this is useful, for example, if you will soon enter a dark theater.

Notice in Fig. 12.8 that there is a region from about 13° to 18° that has neither rods nor cones—the *blind spot*. This is the point at which the optic nerve enters the eye. The blind spot is on the side toward the nose; if an image falls on the blind spot in one eye, it misses the blind spot in the other eye. We are normally not aware of the blind spot, but it is easy to demonstrate. A simple experiment is to place five coins about 10 cm apart in a horizontal row on a table. If you close one eye and look at the central coin with the other, it is possible to adjust your viewing distance to make one of the adjacent coins disappear. If you repeat the experiment with your other eye, the other adjacent coin will disappear. The blind spot is quite large; it covers an angle equal to 11 full moons placed side by side in the sky!

12.3

PROBLEM

At what wavelengths is the eye most sensitive in daylight? At night?

12.4

PROBLEM

What causes the blind spot in the eye? Why are we ordinarily not aware of the blind spot?

12.4 The Threshold of Vision

In 1942, Hecht, Schlaer, and Pirenne published the results of an important experiment on the sensitivity of the rods. The excellent book *Visual Perception* by Cornsweet (see the bibliography at the end of the book) gives an extensive discussion of this experiment, and it is only summarized here. The primary question Hecht et al. posed is: What is the minimum number of photons that will produce the sensation of vision at least 60% of the time? To obtain the minimum number, Hecht et al. had to optimize their experimental conditions. They had to determine (1) the optimum color to use in the test flash, (2) the most sensitive location in the eye, (3) the best diameter to use in the flash, and (4) the best length of time to use in the flash. They obtained the following answers: (1) the rods are the most sensitive at 510 nm; (2) the rods are most abundant at about 20° from the visual axis; (3) the detectability is independent of flash diameters up to 10′ of arc (about the area covered by the degree symbol in 20°), while above this size more light is needed for detection; and (4) for flash times up to about 0.1 s (100 ms) the length of the flash does not affect the detectability, but for longer times more light is necessary.

The final results of their experiment showed that if about 90 photons enter the eye under these optimum conditions, the flash is seen 60% of the time. When these investigators considered all of the light losses in the eye, they estimated that only 10 photons are actually absorbed in the rods. Since the light is distributed over about 350 rods, they felt that it is unlikely that a single rod receives more than one photon. They thus established that a single photon can activate a single rod. Later experiments indicated that as few as 2 photons actually absorbed in the rods can give a visual signal. For comparison, a flashlight with fresh batteries emits about 10^{18} photons each second!

You might wonder why we cannot see a single photon. Several factors are involved, but a fundamental limitation is due to random action potentials that we can call *electrical noise*. The retina is continually generating such noise. Each rod sends a random action potential about every 5 minutes. With 120 million rods in each eye, there are nearly 3 billion random noise pulses per hour. The light signal must be large enough to be seen over this noise. The cones apparently generate even more random pulses and a much larger light signal is needed to be seen in the fovea centralis.

You may have been surprised to learn that if 90 photons enter the eye, 10 or less are actually absorbed in photo receptors. What happens to the others? About 3% are reflected at the surface of the cornea, and about 50% are absorbed in the various structures (cornea, lens, humors). Of those that reach the region of the rods, only about 20% (about 10% of the original number) are absorbed in the rods. The photons that miss the rods are absorbed in the "backstop." Some animals, such as cats, have a reflective coating behind the rods that gives the rods another opportunity to absorb the photons. These animals have eyes that "glow in the dark" if a light shines in them. The reflected light reduces the resolution of small details.

12.5

PROBLEM

If 100 light photons enter the eye, about how many are finally absorbed in light sensors of the retina? [Answer: 10 to 11 photons]

12.5 Diffraction Effects on the Eye

All light waves undergo diffraction as they pass through a small opening. Thus the iris produces a diffraction pattern on the retina (Fig. 12.11). At the normal opening of the pupil (~4 mm) this phenomenon has no practical consequences for our daily vision tasks. However, if the pupil becomes much smaller, for example, 1.0 mm, diffraction

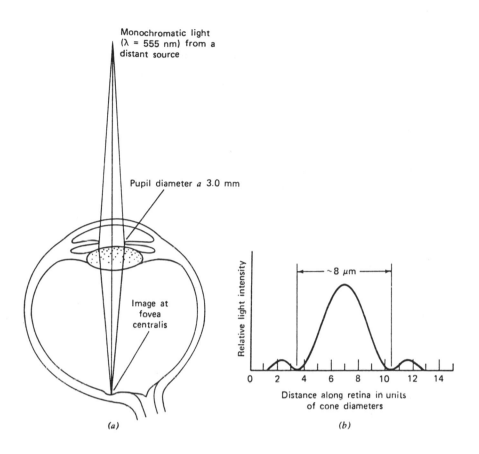

Monochromatic light
(λ = 555 nm) from a
distant source

Pupil diameter *a* 3.0 mm

Image at
fovea
centralis

Relative light intensity

~8 μm

0   2   4   6   8   10   12   14

Distance along retina in units
of cone diameters

(*a*)   (*b*)

Figure 12.11. Diffraction in the eye. (a) Monochromatic light from a distant point source is brought to a focus at the fovea centralis in the retina. (b) The diffraction pattern on the retina produced by a pupil 3.0 mm in diameter consists of a central bright spot 8 μm in diameter surrounded by a ring of light of reduced intensity.

produces a measurable effect on visual acuity. You can demonstrate this effect by reading an eye test chart through a 1.0 mm hole; you should notice a significant decrease in your ability to read the small letters. If your vision is 20/20 (see Section 12.6), looking through a 1 mm hole reduces your vision to 20/60.

All lenses have defects (aberrations). The effect of such aberrations is reduced if the lens opening is made smaller. In the eye, a small pupil improves visual acuity. However, if the pupil is made very small, the acuity becomes worse due to diffraction effects. There is an optimum size for the pupil; best acuity for an emmetropic eye is obtained with a pupil size of 3 to 4 mm—its normal size under good illumination.

A point source of light will not be focused on a single cone because of diffraction effects (Fig. 12.11). The angular spread, 2θ, of the central bright spot at the retina for $\lambda = 555$ nm and a pupil 3.0 mm in diameter a is given by: $2\theta = 2 (1.22)(\lambda/a) = 2(1.22)(555 \times 10^{-9}/3 \times 10^{-3}) = 4.5 \times 10^{-4}$ radians. [Radians are the appropriate SI measure for angles. (radians) = s/r, where r is the radial distance to an object and s is the arc defining the object size, drawn perpendicular to the radius. $180° = \pi$ radians, or $1° = 17.45$ milliradians.]

The diameter of the central bright spot at the retina is the product of the effective aperture to retina distance (17 mm) times 2θ, or $17(4.5 \times 10^{-4}) = 8 \times 10^{-3}$ mm = 8 µm. This spot will include many cones, whose diameters are ~1.1 µm. If the source is bright, for example, a bright star, the next ring of the diffraction pattern can stimulate even more cones. Thus intense point sources of light appear larger than weak point sources, a fact that early astronomers were not aware of when they assigned "magnitudes" to stars; in astronomy this term is now interpreted as an intensity rather than as a size.

12.6

PROBLEM

What is the optimum size of the pupil in an emmetropic eye? Why is this the optimum size? [Answer: 4 mm]

12.6 Visual Acuity—How Sharp Are Your Eyes?

The familiar eye charts used to determine whether we need corrective lenses test the property of our eyes called *visual acuity*. A physicist calls visual acuity the *resolution* of the eyes. In this section we discuss several tests for visual acuity.

The optometrist usually uses a Snellen chart (Fig. 12.12) to test visual acuity. If your eyes test normal at 20/20, it means that you can read detail from 20 ft that a person with good vision can read from 20 ft. [In Europe, where the metric system is used, normal eyes would be reported to test at 6/6, as 6 meters is the standard distance.] If your eyes

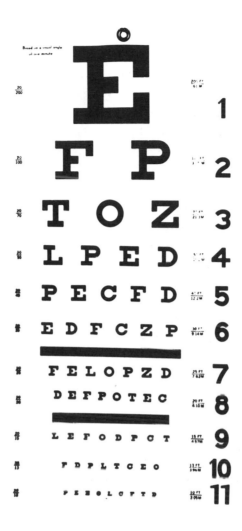

Figure 12.12. A Snellen chart to test vision is usually viewed from 20 ft, or 6 m. The 20/20 line is number 8. In Europe, using the metric system the 20/20 becomes 6/6, meaning you can read line number 8 viewed from 6 m. In this line, the lines that form the letters have an angular width of 1′ of arc at 20 ft. Some letters (e.g., L) are easier to recognize than others (e.g., B and H).

test at 20/40, you can just read from 20 ft the line that a person with good vision can read from 40 ft. The Snellen chart tests many things besides acuity. A person who reads a lot recognizes letters more readily than one who reads little. (You can usually read a line of print with the top or bottom half of it covered.) Some letters, such as A and V, are easy to recognize from their general shape; it is not necessary to distinguish their details. Nevertheless, the Snellen chart is a highly useful tool and will probably continue to be used because of its simplicity.

The visual acuity or resolution of the eye is primarily determined by the characteristics of the cones in the fovea. A common way of testing resolution is to use a pattern of alternating black and white lines that become increasingly narrower. A combination of one white line and one black line is called a line pair (lp). Under optimum conditions, the eye can just barely resolve as separate lines a pattern of about 30 lp/mm; when the eye is twice as far away, it can only resolve 15 lp/mm. The resolution is often given in terms of the angle subtended from the eye. This angle is more or less independent of viewing distance. The minimum angle between two black lines that can be seen as separate is about 0.3 milliradians. To be seen as separate, the two lines must fall on alternate rows of cones so that the cones between will see the white strip (Fig. 12.13). The smallest black dot that you can see under optimum conditions is 2.3×10^{-6} radian. The resolution rapidly gets worse as the image moves away from the fovea centralis. At 0.175 radian (10°) from the fovea the acuity is worse by a factor of 10. If the lighting is not optimum, the resolution also deteriorates (Fig. 12.14).

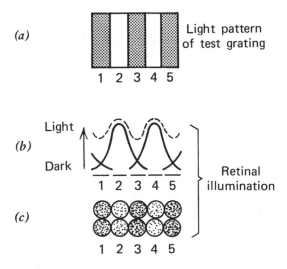

Figure 12.13. To see closely spaced lines (a) as separate, it is necessary for the images of the lines to fall on alternate rows of cones as shown schematically in b. All the cones receive some light (c), but more falls on cones corresponding to the white strips. (Adapted from Ruch, T. C. in T. C. Ruch and H. D. Patton, Eds., *Physiology and Biophysics*, 19th Ed., © W. B. Saunders Company, Philadelphia, 1965, p. 428.

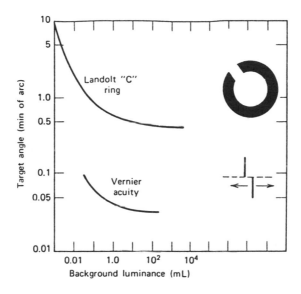

Figure 12.14. Visual acuity improves with better lighting. The top curve shows the acuity of the Landolt C ring, where the gap direction must be recognized. Vernier acuity (the ability to align two lines) is much better than acuity for the Landolt C. The typical brightness level for reading is about 100 millilamberts (mL).

One test of acuity used often by scientists is aligning ends of two parallel lines so that they appear to be a single continuous line (Fig. 12.14). Alignment such as this is involved when using a measuring instrument such as a vernier scale, and it is referred to as vernier acuity. It is possible for a trained person to align two fine lines under optimum conditions to within 9×10^{-6} radians, much less than the 3×10^{-4} radians needed to resolve two lines.

The resolution of white letters on a black background is about 10^{-3} radians, while that of black letters on a white background is 3×10^{-4} radians. This means that a good eye would read only 20/60 on the Snellen chart if the chart were made in the white-on-black format! This fact is of practical importance in making projection slides for lectures; light letters on a dark background are not as easy to read as the conventional dark letters on a light background.

The ability of the eye to recognize separate lines also depends on the relative "blackness" and "whiteness" of the lines. Resolution is much worse if the lines are two slightly different shades of gray than if one is black and one is white. The contrast C between two areas is defined as $C = (I_1 - I_2)/(I_1 + I_2)$, where I_1 and I_2 are the light intensities from the two areas. The low contrast between two areas of interest on an x-ray often severely limits the usefulness of the x-ray image.

Before discussing contrast further, we need to define a unit for measuring the density, "darkness" or opaqueness, of an x-ray film or any other optical absorber. The *optical density* OD is defined as $OD = \log_{10}(I_0/I)$ where I_0 is the light intensity without the absorber and I is the intensity with the absorber. For example, a piece of film that transmits 10% of the incident light has an optical density of $\log_{10}(1/0.1) = \log_{10}10 = 1.0$. A film that absorbs 99% of the light has an $I_0/I = 100$, and the $OD = \log_{10}(10^2) = 2.0$. An OD = 3 means that only 0.1% of the light is transmitted. A light must be bright to be seen through an OD = 3 filter.

From the properties of logarithms, ODs are additive. Therefore, you can put together two neutral density filters of OD = 1.0 (available from most camera stores) to get OD = 2.0, you can put three together to get OD = 3.0, and so forth. Even a "perfectly clear" piece of film has a small optical density due to the reflection at the surfaces. Typically, about 3% is reflected back from a clean optical surface. The range of optical densities used for most x-ray viewing is 0.3 to 2.0. However, darker areas of as much as 3.0 can be viewed with a spotlight. Taking x-rays of the optimum darkness is important for obtaining the maximum amount of medical information.

If two films are placed adjacent to each other, there must be a difference in their optical densities for the eyes to recognize them as being different. This difference depends on the light intensity (Fig. 12.15). At very low light levels, when we are using our rods, a factor-of-2 difference in light may be required. Under optimum light levels a 1% or 2% difference is detectable. Usually an x-ray does not have a nice sharp border where two areas meet; the optical density gradually changes from one area to another. As much as a 20% change in light intensity may then be required for two films of different optical densities to be recognized as different, even under optimum conditions.

12.7

PROBLEM

What is the fractional amount of light transmitted by a film of OD = 0.5? [Answer: About 1/3, or 33%]

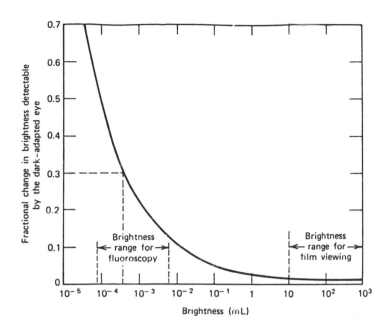

Figure 12.15. The contrast needed to distinguish different shades of gray at different light levels. At levels used for conventional fluoroscopy, a 30% change in light is needed to be detectable. (Adapted from H. E. Johns, *The Physics of Radiology*, 2nd Ed., C. C. Thomas, Springfield, Ill., 1961, p. 597.)

12.8

PROBLEM

If the structures in the eye absorb 50% of the photons before they reach the retina, what is the OD of the eye? [Answer: O.D. = 0.3]

12.7 Optical Illusions and Related Phenomena

Viewing shades of gray plays tricks on the mind. Figure 12.16a shows strips of uniform shades of gray next to each other. Rather than looking uniform, a given bar looks darker where it borders a lighter bar and lighter on the side near a darker bar. (To help see that a bar is uniform, use white paper to cover the adjacent bars.) This effect is produced by

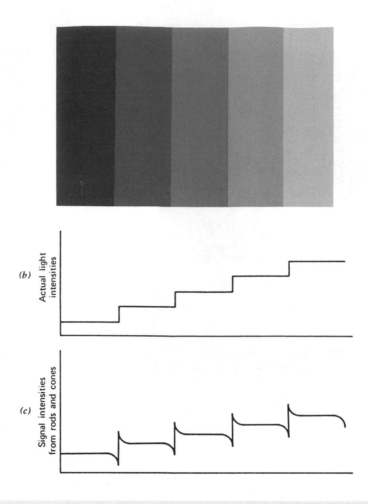

Figure 12.16. Strips of gray (a) that are uniformly shaded as shown by their actual light intensities (b) are perceived by the rods and cones as darker near a light bar and lighter near a dark bar (Mach bands) (c).

interactions between adjacent groups of light sensors in the eye as illustrated schematically in Fig. 12.16c. This "edge enhancement" would appear to help radiologists see borders on x-ray films. Unfortunately, it helps only where they do not need the help—where the contrast is already good. The effect disappears when the contrast is small.

Another optical illusion is shown in Fig. 12.17. The circles in this figure are all the same shade of gray, but the circle surrounded by a light area looks darker. This effect has practical significance for viewing x-ray films on an illuminator. If the film does not cover the entire light source, the bright light around the film makes it appear darker.

Figure 12.17. The perception of the darkness of an image depends on the darkness of its background. All three circles are the same shade of gray.

Like many of the other sensory nerves, the nerve cells in the eye stop sending signals in the presence of a steady stimulus. The eye overcomes this handicap by continuously jiggling during normal use in addition to making gross movements. For a simple demonstration of the fading of the signal from the eye, find a nice large plain ceiling or wall that has a noticeable border between two sections. If you stare steadily at this area, you will see the two sections gradually take on the same shade and the border between them disappear. If you shift your eyes, the border will return!

If you look into the eye with an ophthalmoscope you can see the many blood vessels in the retina (Fig. 12.18). These blood vessels block out light to the rods and cones behind them. The reason we do not normally see these vessels is that the shadows they produce are always over the same rods and cones and this steady signal fades in moments after we open our eyes in the morning. Most people are not prepared to look at the blood vessels in their eyes first thing in the morning anyway! You can, however, see these vessels by transillumination with a penlight. With your eye closed, hold a penlight against your eyelid and move the penlight rapidly back and forth. Some of the light will penetrate the eyelid and sclera and cause the blood vessels to cast shadows on different rods and cones, producing a visible image of the network of blood vessels. This technique is used as a clinical test to ascertain the presence of retinal function in eyes in which the retina cannot be seen.

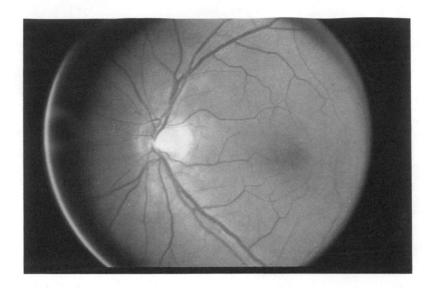

Figure 12.18. The blood vessels in the retina. These blood vessels are not seen during normal vision because their shadows always fall on the same rods and cones and the signal fades rapidly.

Everyone has at one time or another appeared to see light with the eyes closed. These "lights" are called *phosphenes* and can be stimulated by pressing on the eye with the fingers or by closing the eyes very tightly. Phosphenes are produced by the stimulation of some of the normal light sensors. The brain interprets any signals received from the optic nerve as light; it cannot distinguish the various sources of signals from the optic nerve. If you receive a blow on the head, this will stimulate some of the light sensing nerves and you will "see stars." *Electrophosphenes* can be produced if a small voltage (~4 V) is placed across your eye when your eyes are closed and dark adapted; you will see "light" each time the voltage is turned on or off. Since the nerves normally transmit signals of less than 0.1 V, it is not surprising that the rapid changes produced by this technique trigger some nerve pulses. A changing magnetic field near the eye produces magnetophosphenes.

We will not discuss in detail the important relationships between the brain and the eyes. Everybody has seen illustrations in which straight lines look curved and lines of equal length appear to be quite different in length (Fig. 12.19). When you move your eyes or your head, the image in your eyes sweeps rapidly; you might think the brain would get the impression that the room is in motion. Apparently as the brain signals

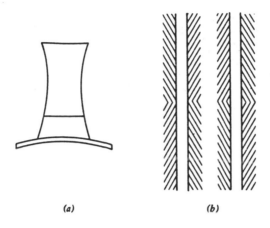

Figure 12.19. (a) The width-height optical illusion. The base is as long as the height. (b) The straight-line optical illusion. The surrounding patterns make the straight lines appear curved.

the muscles to move the eyes or head, it also informs the visual cortex, and we are not confused by the illusion of a moving room. However, if the eye moves by some external force it can give the impression that the room is moving. Close one eye and half close the other. Push with a finger on the eyelid of the half-closed eye and you will see the image you are viewing move; no signals have been sent to the visual cortex to cancel out the image movement produced by your finger pressure.

A characteristic of the eye-brain system that we take for granted is the ability of the brain to fuse the slightly different images from the eyes into a three-dimensional (3-D) image. An artificial 3-D effect can be produced by showing slightly different pictures to the two eyes. About 150 years ago Wheatstone invented the stereoscope, which is still the basic device used to view stereoscopic x-rays. For stereoscopic x-rays, two x-rays are taken of the same part of the body from slightly different angles corresponding to the views normally seen by the two eyes. The two x-rays are then placed in a stereoscope so that each eye sees one image; these images are fused by the brain into a 3-D image (Fig. 12.20). Stereoscopic x-rays are often taken of the skull.

The brain merges the signals from both eyes even if one is badly out of focus or if the image in one eye has a magnification as much as 5% larger than that in the other eye! For a simple demonstration of the brain's ability to merge two quite different images, roll up a tube of paper, hold it to one eye, and look at a plain well-lit wall. Hold your

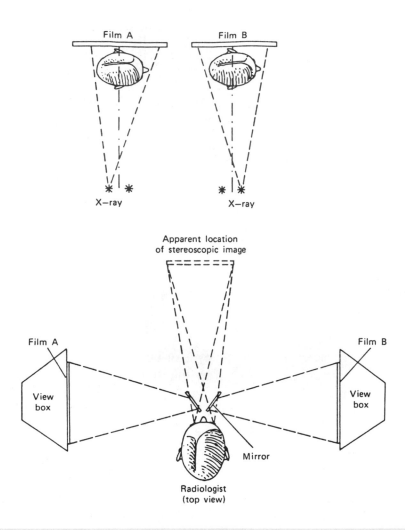

Figure 12.20. The use of a stereoscopic system to obtain a three-dimensional view. The x-rays are taken from slightly different angles to simulate the spacing of the eyes.

hand alongside the paper tube and look at it with the other eye. You will be surprised (we hope) to see a hole in your hand (Fig. 12.21)!

Sometimes you think you see something because you expect to see it. If a radiologist makes a medical diagnosis on the basis of such an imagined image, this type of illusion can be dangerous. This situation is summed up in the motto, "If I hadn't believed it with my own mind, I never would have seen it!"

When you view a flash of light, the visual image of the light persists for some time after the flash. That is, after the flash there is a period

Figure 12.21. By looking through a tube with both eyes open as shown, you can painlessly put a hole in your hand.

of many milliseconds when the brain thinks the light is still on. If the frequency of consecutive flashes is increased, at some rate the eye-brain system no longer recognizes the light as flashing—*flicker-fusion* is said to exist. This rate depends on the intensity of the flashes; bright flashes may not fuse into a "steady light " until about 50 Hz, whereas dim flashes will appear as a steady light at as low as 12 Hz. The rods have a higher flicker-fusion rate than the cones. A flicker in your peripheral vision may fuse when you look at it directly.

The ability of the eye to fuse flashing lights into "steady" light is the basis of movies, in which 16 to 32 images per second are flashed on the screen. We are not aware that the screen is blank most of the time.

12.9

PROBLEM

Why do you normally not see the blood vessels in your retina?

12.8 Defective Vision and Its Correction

Glasses (corrective lenses) to help defective vision were among the first prosthetic devices invented. It is not clear who gets credit for the invention, but the first reference to the use of glasses (hand held) dates from around 1300. About a century later someone figured out how to attach glasses to the head. However, wearing glasses in public was considered in bad taste even in the 1800s. Until the 20th century, their use was restricted to the well-to-do. The history of corrective lenses includes the invention of the curved lens to reduce aberrations when looking at an angle; bifocals, primarily for old people; and contact lenses, primarily for young people.

In order to discuss the strength of a corrective lens for a defective eye we need to review the basic equations of simple lenses. There is a simple relationship between the focal length F, the object distance P, and the image distance Q of a thin lens (Fig. 12.22)

$$1/F = 1/P + 1/Q. \tag{12.1}$$

If F is measured in meters, then $1/F$ is the *lens strength* in diopters (D). The focal length of a converging lens is defined to be positive. A lens with a focal length of $+0.1$ m has a strength of $+10$ D. The focal length F of a diverging lens is defined to be negative. A negative lens with a focal length of -0.5 m has a strength of -2 D.

The focal length F of a combination of two lenses, focal lengths F_1 and F_2, which are considered to be essentially in contact with each other, is given by $(1/F) = (1/F_1) + (1/F_2)$. Similarly, for three lenses the equation is $(1/F_{comb}) = (1/F_1) + (1/F_2) + (1/F_3)$. Unless you like working with fractions the solution for F is rather tedious. However, you will note that if each of the focal lengths is given in meters, the equation says that the strength of the combination in diopters is equal to the sum of the diopters of the various lenses, i.e., $D_{comb} = D_1 + D_2 + D_3$.

Let us assume the image distance from the cornea and lens of the eye to the retina is $Q = 0.02$ m (0.017 m is a more correct value, but the arithmetic is harder). When the normal eye is focused at a great distance (infinity), the focal length F of the eye is the same as Q, as we see from

$$D_{far} = 1/F_{far} = 1/\infty + 1/Q = 0 + 1/0.02 \text{ m} = 50 \text{ D}. \tag{12.2}$$

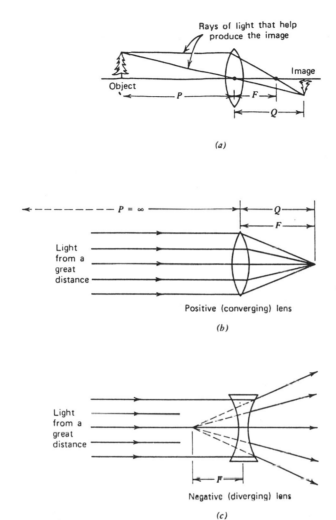

Rays of light that help produce the image

Image

Object

P

F

Q

(a)

$P = \infty$

Q

F

Light from a great distance

Positive (converging) lens

(b)

Light from a great distance

F

Negative (diverging) lens

(c)

Figure 12.22. (a) The distance from the lens to the object P and the distance from the lens to the place where the image is formed Q are related to the focal length F of a positive lens by the equation $1/P + 1/Q = 1/F$. (b) Light coming from a great distance to a positive lens converges at the focus of the lens; the image distance Q is equal to the focal length F. (c) Light from a great distance striking a negative lens diverges. The light appears to diverge from the focal point on the left side of the lens. No image is formed.

We say that the eye, viewing an object at a great distance, has a strength of 50 D. If the eye next focuses on a near object, say at $P = 0.25$ m (~10 in.), then

$$D_{near} = 1/F_{near} = (1/0.25) + (1/0.02) = 4 + 50 = 54 \text{ D}. \quad (12.3)$$

We say that the eye, for this near object, has a strength of 54 D. For vision to be good at both large distances (e.g., far = ∞) and close up (e.g., near = 0.25 m), the eye must have an accommodation of at least

$$\text{necessary accommodation} = D_{near} - D_{far} = 54\,D - 50\,D = 4\,D. \qquad (12.4)$$

Now let us discuss defective eyesight due to focusing (refractive) problems—*ametropia*. Ametropia affects over half of the population of the United States. It is often possible to correct it completely with glasses or by the use of laser surgery to change the shape of the cornea. There are four general types of ametropia: *myopia* (near-sightedness), *hyperopia* or *hypermetropia* (far-sightedness), *astigmatism* (asymmetrical focusing), and *presbyopia* (old sight) or lack of accommodation. Figure 12.23 illustrates these conditions schematically and shows the regions where blurring occurs. For each eye we define the near point as the closest distance at which it can see clearly; the far point is the greatest distance at which it has good vision. The various focusing problems and their characteristics are summarized in Table 12.2.

Table 12.2. A Summary of Various Focusing Problems and Their Characteristics

Focusing Problem	Common Name	Usual Cause	Corrected With[*]
Myopia	Near-sightedness	Long eyeball or cornea too curved	Negative lens
Hyperopia	Far-sightedness	Short eyeball or cornea not curved enough	Positive lens
Astigmatism	—	Unequal curvature of cornea	Cylindrical lens or hard contact lens
Presbyopia	Old-age vision	Lack of accommodation	Bifocals or trifocals

[*]All except presbyopia can be corrected with laser surgery.

The myopic individual usually has too long an eyeball or too much curvature of the cornea; distant objects come to a focus in front of the retina, and the rays diverge to cause a blurred image at the retina (Fig. 12.24b). This condition is easily corrected with a negative lens.

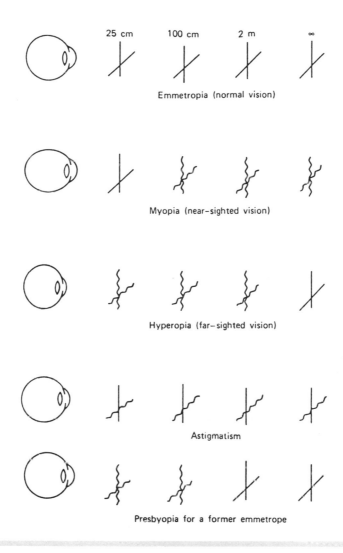

Figure 12.23. Schematic of normal and defective focusing. Wavy lines indicate a blurred image on the retina.

PROBLEM 12.10

Determine the strength of a lens needed to correct a myopic eye which has a far point of P = 1.0 m (far point means the greatest distance for which an object can be imaged on the retina). As above, take the image (lens to retina) distance to be Q = 0.02 m.

[Answer: A negative lens of −1.0 D will correct this vision.]

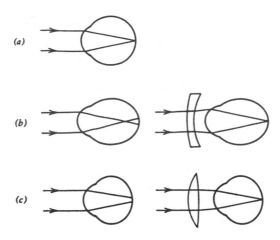

Figure 12.24. Focusing properties of the eye. (a) The normal or emmetropic eye focuses the image on the retina. (b) The near-sighted or myopic eye focuses the image in front of the retina. This problem is corrected with a negative lens. (c) The far-sighted or hyperopic eye focuses the image behind the retina. This problem is corrected with a positive lens.

A hyperopic eye has a near point further away than normal and uses some of its accommodation to see distant objects clearly. The usual cause of hyperopia is too short an eyeball (Fig. 12.24c). A positive lens is used to correct this condition.

12.11

PROBLEM

Determine the strength of a lens needed to allow a person with a far-sighted eye with a near point of P_{near} = 2.0 m (near point means the closest distance for which an object can be imaged on the retina) to read comfortably at 0.25 m? [Answer: A corrective lens of 54 – 50.5 = +3.5 D would be prescribed for this eye.]

You can check to see if you are myopic or hyperopic. Look through a pinhole at a well-illuminated distinct object, for example, a streetlight (Fig. 12.25). Move the pinhole up and down in front of your eye. If you are an emmetrope, you will not see any motion of the image; if you are myopic, the image on the retina will move in the direction opposite that of the card and will be interpreted by the brain as moving in the same direction; and if you are hyperopic, the motion of the image on the retina

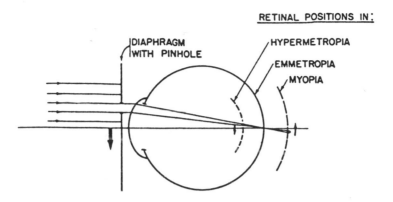

Figure 12.25. Moving a pinhole in front of the ametropic eye while it is looking at a distant object causes apparent motion of the object. Emmetropes see no motion.

will be in the same direction as the card and will appear to be moving in the opposite direction.

You can also easily check whether glasses have positive or negative lenses by looking at an object through one lens held some distance away. When you move the lens, the object also appears to move. If it moves in the same direction as the motion of the lens, it is a negative lens; if it moves in the opposite direction, it is a positive lens. Another test is to hold the lens over some printing. If it enlarges the printing, the lens is positive; if it makes the printing smaller, the lens is negative.

In astigmatism, the curvature of the cornea is uneven. Astigmatism cannot be corrected by a simple positive or negative lens. A simple test for astigmatism is to look at a pattern of radial lines (Fig. 12.26). An astigmatic eye will see lines going in one direction more clearly than lines going in other directions. Astigmatism is corrected with an asymmetric lens in which the strength is greater in one direction than in the perpendicular direction (Fig. 12.27). If you wear glasses to correct astigmatism, hold your glasses some distance from your head and rotate them while looking at an object through one lens. You will notice that the object appears to change shape as you rotate the glasses.

Often a person older than 50 has trouble reading fine print; when a book is held far enough away to focus clearly, the print is too small, making it impossible to distinguish the letters. Although reading in a bright light helps because it narrows the pupil and thus provides a better depth of focus, this individual will need reading glasses. If glasses are already being worn to correct a vision defect, bifocals, or even

Figure 12.26. A simple test for astigmatism. An eye with astigmatism sees lines going in one direction more clearly than lines going in the other directions.

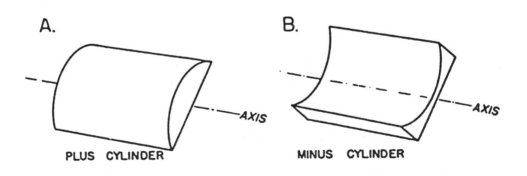

Figure 12.27. Astigmatism is corrected by adding a cylindrical lens to a spherical lens. The cylinder may be either (A) converging (plus cylinder) or (B) diverging (minus cylinder). (From M. L. Rubin, *Optics for Clinicians*, Triad Scientific Publishing, Gainesville, FL, 1971, p. 94.)

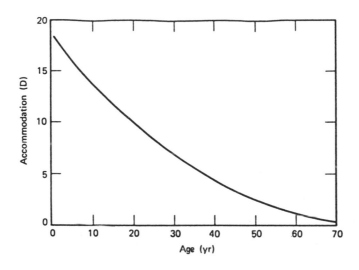

Figure 12.28. Loss of accommodation with age. The decrease in accommodation usually becomes noticeable after age 40.

trifocals, will be necessary. This problem is the result of the loss of accommodation with age (Fig. 12.28). The lens becomes less pliable, and when the tension on it is released, its dimensions change only slightly. As you can see in Fig. 12.28, this loss of accommodation starts at an early age. (See Problem 12.11 for calculations of the lens strength needed for presbyopic eyes.)

If you wear corrective lenses, you should carry a copy of your prescription. If you lose your glasses far from home, you can have a new pair made without a re-examination. One prescription for glasses reads as follows:

	Sphere	Cylinder		Axis	Add
O.D.	−1.25	−1.25	×	180	+1.25
O.S.	−1.75	−1.75	×	163	+1.25

This means that the right eye (O.D.) needs a spherical lens of −1.25 D added to a cylindrical lens of −1.25 D in the horizontal plane (180°). In the reading portion of the bifocal, a spherical lens of +1.25 D is added to the above prescription. That is, the effective strength in the lower portion of the right lens is a cylindrical lens of −1.25 D to correct the astigmatism. The prescription for the left eye (O.S.) is interpreted in the same way.

Often fractions such as 20/20 or 6/6 are included after the prescription for each eye. This indicates that when the glasses are worn, the eye will be able to read 20/20; that is, to read at 20 ft the line on the Snellen chart that a normal eye can read at 20 ft. The ratio 6/6 is the same statement in meters, since 6 m is about 20 ft.

The idea of *contact lenses* dates back to before 1900, but not until the late 1950s were many of the technical and medical problems associated with them satisfactorily solved. Contact lenses are made of either hard (gas permeable) or soft plastic. The gas permeable contact lens rests on a film of tears on the very front (apex) of the cornea, one of the most sensitive spots on the body.

From a physics standpoint the contact lens performs the same function as an ordinary corrective lens. However, the focal power of the combination of two lenses depends on the separation between them. The separation between ordinary glasses and the cornea is fairly well determined by the construction of the frame. Changing this separation does not have much effect unless the glasses have very strong lenses. If you wear glasses, move them a few centimeters from your eyes and see if you can notice much effect. The direct contact between a contact lens and the cornea, however, noticeably affects the prescription. A myopic person switching from glasses to contact lenses would need weaker negative lenses, and a hyperope would need stronger positive lenses. The image size on the retina produced by contact lenses is different from that produced by ordinary glasses; it is larger in myopia and smaller in hyperopia. Wearing contact lenses requires more accommodative effort for myopia and less for hyperopia. Thus a myopic person in the early stages of presbyopia is hindered by contact lenses, and the hyperope in the same situation is helped.

Since a contact lens does not have a fixed orientation, the question of correcting for astigmatism must be resolved. You will recall that astigmatism is caused by nonuniform curvature of the cornea. The hard contact lens is made to fit the largest radius of the cornea. The empty space under the rest of the cornea is filled with tear fluid. The tears correct for astigmatism by making a symmetrical "new cornea."

Developed in Czechoslovakia in the 1950s, soft plastic contact lenses are much more comfortable to wear than hard plastic lenses. New users can often wear them all day long, even on the first day. Also, soft lenses are permeable to gases so that oxygen can reach the cornea directly. With hard plastic lenses, the oxygen is dissolved in the tear layer, and each blink carries a fresh supply under the lens to the cornea. The main disadvantages of soft contact lenses are: (1) they cost more; and (2) since they conform to the cornea, they do not correct for astigmatism. Soft contact lenses used to require a special daily cleaning procedure. Now, the most popular style is the continuous wear or soft, disposable contacts. There are two types: one type is worn for only a day and then discarded; the other type is worn for two weeks, if you take them out each night. If you want to wear them continuously, day and night, you must take them out after a week and clean them and discard them at the end of the second week. The major drawback of contact lenses, compared to ordinary glasses, is the increased probability of an eye infection.

Contact lenses are sometimes used for other than cosmetic reasons. A patient with a cornea transplant often has complex astigmatism as a result of the stitches. Hard contact lenses are needed to remove the distortion caused by the stitches. In rough sports such as football and soccer, players often wear contact lenses for safety reasons. Soft plastic lenses are also used to administer medication directly to the cornea over a period of hours. The lens is soaked in the medication before it is put in the eye. Readers wishing to learn more about contact lenses should consult the bibliography at the end of the book.

12.12

PROBLEM

What is presbyopia? What is meant by the term "accommodation"?

12.13

PROBLEM

What is meant by 20/40 vision?

12.14 **PROBLEM** On examination, an optometrist finds that a patient who formerly had good vision now has a near point of 0.5 m and likes to read at a distance of 0.25 m. What is this patient's accommodation, and what strength reading glasses should be prescribed? (Hint: Remember that an eye focused at infinity has a strength of 50 D.)
[Answer: Reading glasses which have a strength +2 D should be prescribed. These can be purchased inexpensively at a local drug store.]

12.15 **PROBLEM** If a person with myopia has a near point of 15 cm without glasses and wears a corrective lens of −1.0 D, what is the near point when the person is wearing glasses? [Answer: 0.18 m, or 18 cm]

12.16 **PROBLEM** If an emmetrope has an accommodation of 3 D, what is that person's near point? [Answer: 0.33 m, or 33 cm]

12.17 **PROBLEM** If a former emmetrope wears reading glasses of +2 D to read at a distance of 25 cm, what is the near point without glasses?
[Answer: 0.5 m, or 50 cm]

12.18 **PROBLEM** How do hard contact lenses correct astigmatism? Why does this correction not occur when soft contact lenses are worn?

12.9 Color Vision and Chromatic Aberration

One of the remarkable abilities of the eye is its ability to see color. The exact mechanism of color vision is not well understood, but it is fairly well accepted that there are three types of cones that respond to light from three different parts of the spectrum. Images on a color TV are produced by a method similar in some aspects. If you examine a color TV screen with a magnifying glass, you will see a myriad of small red, green, and blue dots. These dots can produce, in different combinations, all of the colors of the spectrum. It is thought that in an analogous way, signals are sent to the brain from the three "colored" cones in different combinations, permitting the brain to determine color.

If one of the color sets is gone, a person is color blind, meaning that certain colors are confused. Approximately 8% of all men and 0.5% of all women have some type of color blindness. It is rare that someone is completely color blind; that is, sees only shades of gray.

Chromatic aberration is a common defect in simple lenses caused by the change of the index of refraction with wavelength. (This change in the index of refraction with wavelengths permits a prism to separate white light into a rainbow of colors.) Chromatic aberration causes different colors to come to a focus at different distances. In a simple lens this will produce colored fringes on the image of a white object. Chromatic aberration is greatly reduced in an expensive camera because a lens is used that is made of several types of glass chosen to compensate for the change of the index of refraction with wavelength.

The eyes do not produce colored fringes, or if they do, we are unaware of them. Nevertheless, the acuity of the eye is affected by the different focal lengths for the different colors. For example, the change in focal length from deep blue (390 nm) to deep red (760 nm) is almost 0.7 mm, or more than twice the thickness of the retina! It takes about 2.5 D of lens power to shift the focal length by that amount (Fig. 12.29). Chromatic aberration in the eye can be demonstrated by the simple arrangement illustrated in Fig. 12.30. Look at the red filament of a clear light bulb through a thick cobalt glass (blue) filter. You will see two images of the filament, one red and one blue, next to each other. If your eye could focus red and blue equally, the two images would be superimposed. One of the reasons chromatic aberration is no problem in ordinary vision is that it is rare to encounter such an extreme situation of colors from the ends of the spectrum only. The eyes have their best sensitivity in a rather narrow yellow band in the center of the visible spectrum (Fig. 12.29), and the iris limits the light to the center of the lens,

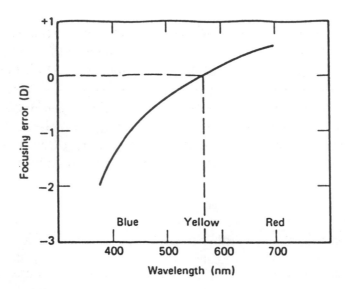

Figure 12.29. Different colors are focused differently by the eye. There is about 2.5 D difference in focusing power between deep blue and deep red. The graph shows this effect for a normal eye (normalized to yellow light).

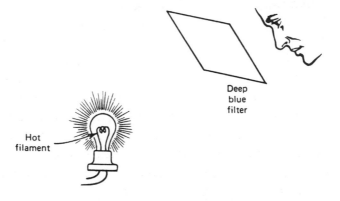

Figure 12.30. A simple arrangement to demonstrate chromatic aberration of the eye. The eye sees separate red and blue filaments due to the different focal properties of the eye for the ends of the spectrum.

where chromatic aberration is least. The yellowish adult lens acts as a filter to remove some of the reds and blues from the light that strikes the retina, although no observable improvement is made by wearing yellow-tinted glasses. Monochromatic illumination plus a corrective lens gives higher visual acuity than white light. The best acuity is obtained with yellow light, although the effect is not very dependent on wavelength.

A special color effect that is sometimes noticeable at dusk is called the *Purkinje effect*. Purkinje noticed that at dusk the blue blossoms on his flowers appeared more brilliant than the red blossoms. This effect is caused by the shift of the best sensitivity of the eyes toward the blue as the rods take over from the cones at low light levels. Since the eyes and corrective lenses are optimized for yellow light, this shift toward the blue produces a refractive error of about 1.0 D. In other words, for night vision you should wear glasses with an additional −1.0 D!

12.10 Instruments Used in Ophthalmology

There are three principal instruments used to examine the eye; the *ophthalmoscope*, which permits the physician to examine the interior of the eye; the *retinoscope*, which measures the focusing power of the eye; and the *keratometer*, which measures the curvature of the cornea. Another instrument, the *tonometer*, measures the pressure in the eye. The *lensometer* is not used to study the eye: it determines the prescription of an unknown lens.

The *ophthalmoscope* is by far the most used, and several versions have been designed. It was invented in 1851 by Helmholtz, an early "medical physicist." The principle of the ophthalmoscope is shown in Fig. 12.31. Bright light is projected into the subject's eye, and the returning light from the subject's retina is positioned so that it can be focused by the examiner. The lens system of the patient's eye acts as a built-in magnifier. A trained individual can detect more than eye problems with an ophthalmoscope since increased pressure inside the skull (for example, due to a brain tumor) can cause a noticeable change in the interior of the eye (papilledema).

The *retinoscope* is used to determine the prescription of a corrective lens without the patient's active participation, although the eye has to be open and in a position suitable for examination. This technique can be used, for example, on an anesthetized infant. The retinoscope is also sometimes used to check the prescription determined by the usual "which is better, the first or the second" technique.

A streak of light from the retinoscope is projected into the patient's unaccommodated dilated eye. This streak of light is reflected from the retina and acts as a light source for the operator. The retina's function in retinoscopy is the reverse of its normal function (Fig. 12.32a). Since an object at the eye's far point would be focused at the retina of a relaxed

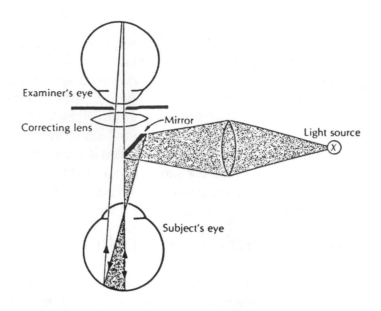

Figure 12.31. The opthalmoscope permits the examination of the retina. Light is directed into the patient's eye to permit the examiner to view the retina through a correcting lens.

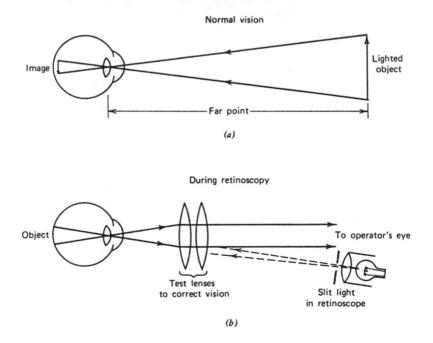

Figure 12.32. (a) The eye during normal vision. (b) During retinoscopy, reflected light from the patient's retina acts as the object. Lenses are added in front of the eye to focus the image from the retina at the operator's eye.

eye, a light from the retina of a relaxed eye will produce a focused image at the far point. The operator views the patient's eye through the retinoscope and adds lenses in front of the patient's eye (positive or negative, as needed) to cause the image from the patient's retina to be focused at the operator's own eye (Fig. 12.32b). To determine the prescription needed to correct the patient's eye, the operator must change the lens power of these added lenses by the dioptric power needed to focus the same eye at the "operator distance." If the operator distance is 0.67 m, then –1.5 D must be added to obtain the correction to focus at infinity.

The *keratometer* is an instrument that measures the curvature of the cornea. This measurement is needed to fit contact lenses. If we illuminate an object of known size, placed a known distance from a convex mirror, and measure the size of the reflected image, we can determine the curvature of the mirror. In keratometry the cornea acts as a convex mirror. The reflected image is located at the focal plane, a distance r/2 behind the surface of the cornea (Fig. 12.33). The keratometer produces a lighted circle that is reflected from the cornea while the patient's head is held in a fixed position. The operator adjusts a focus control to place the instrument a known distance from the cornea (Fig. 12.34). Part of the reflected image passes through a prism that causes a second image to be seen by the operator. The operator determines the size of the reflected image by adjusting the angle of the prism to produce a coincidence of marker lines in the two images.

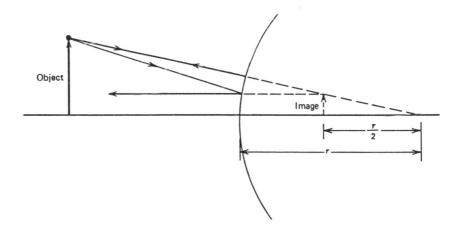

Figure 12.33. The reflected image from the cornea is located r/2 behind the surface, where r is the radius of curvature of the cornea. This fact is used to measure the curvature of the cornea with the keratometer.

Figure 12.34. The keratometer is an instrument used to measure the curvature of the cornea. This measurement is necessary for prescribing contact lenses. (Courtesy of Bausch & Lomb, Rochester, NY.)

The position of the prism after this adjustment is indicated on a dial that is calibrated in diopters of focusing power of the cornea. The average value is 44 D, which corresponds to a cornea with a radius of curvature of 7.7 mm. Because astigmatism is common, the measurement of the image size is made on the long axis of the "cylindrical" lens of the eye and at right angles to it. The curvature of the contact lens is made to fit the larger radius.

A simple physics experiment involves determining the focal length of a lens. If you have a positive lens (e.g., a magnifying glass), you can produce a focused image of a distant object (e.g., the sun). The image will be at the focal point of the lens. You can measure the distance from the lens to the image to determine the focal length. This technique will not work for a negative lens since no real image will be formed, but a simple modification will permit you to use the same idea. You can combine a negative lens with a strong positive lens of known strength and then produce an image of a distant object to get the focal length of the combination. From this, you can determine the dioptric strength of the two lenses. You can then determine the diopters of the negative lens from

$$D_x + D_{known} = D_{measured}.$$

These techniques for measuring the power of a lens are suitable for a physics laboratory, but they are clumsy and inconvenient for an ophthalmologist, optometrist, or optician. For routine use, a commercial *lensometer* is more convenient (Fig. 12.35). It moves an illuminated object until it is at the focal point of a lens combination consisting of a fixed positive field lens and the unknown lens. The parallel rays emerging from the lenses are viewed by a telescope focused at infinity. The fixed field lens is placed a distance equal to its focal length from the unknown lens. This placement conveniently makes the position of the movable lighted object a linear function of the strength of the unknown lens. That is, the scale of diopters (Fig. 12.35) is uniformly divided. When the lighted object is at the focal point of the field lens, the lensometer reads zero D. As it moves further away from the field lens, the lensometer reads in negative diopters; and as it moves closer, the lensometer reads in positive diopters. For a cylindrical lens (used to correct astigmatism) the strength of each lens axis is measured separately and the angle of the cylindrical lens can be determined.

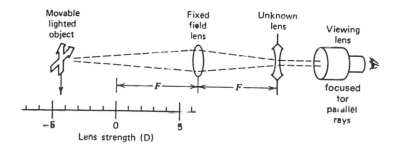

Figure 12.35. A lensometer measures the power of an unknown lens by moving an illuminated object until a sharp image is seen in the viewing lens. The viewing lens is focused for parallel light. In this schematic a −4 D lens is being measured.

It has been known since before 1900 that high eye pressure is related to the condition of *glaucoma*. This disease narrows the field of view (produces "tunnel vision") and leads to blindness if not treated. If the canal of Schlem is too narrow, it requires more pressure for the aqueous humor to flow out. The fluids in the eyeball are normally under a pressure of 1.6 to 3.0 kPa (12 to 23 mm Hg); in glaucoma, the pressure may go up to 11 kPa (85 mm Hg)—comparable to the blood pressure.

In about 1900 in Germany, Schiotz invented an instrument for measuring the intraocular pressure the Schiotz tonometer. The basic technique called for resting the tonometer on the anesthetized cornea with the patient supine (lying face up). The center plunger causes a slight depression in the cornea (Fig. 12.36a). The position of the plunger indicates on a scale the internal pressure in the eye. The force of the plunger could be varied by adding various weights. The standard weights had masses of 5.5, 7.5, 10.0, and 15.0 g. The plunger alone has a mass of 11.0 g. With the standard 5.5 g mass, there is 16.5 g resting on a small area of the cornea (Fig. 12.36b). This increases the internal pressure about 2 kPa (15 mm Hg) depending on the rigidity of the eye.

The pressure measured by the tonometer is the original pressure plus the increase due to the instrument. To remove the effect of the rigidity of the eye, another measurement is taken with a heavier weight or with the Goldmann tonometer (described below). The two readings permit the operator to determine, with the help of tables, the original pressure and the rigidity of the eye. The rigidity has no known diagnostic value.

The Schiotz tonometer was modified about 1950 to give readings electronically. A coil magnetically sensed the position of the plunger (Fig. 12.37). One advantage of this model was that it recorded the change of pressure with time. Figure 12.38 shows such a record, called a *tonograph*. Note the fluctuation in pressure due to the pulse in the arteries. The decrease in pressure indicates that the aqueous fluid is leaving the eye faster than normal under the pressure produced by the tonometer. The outflow can be estimated from the slope of the tonograph. The outflow is normally 2 to 6 ml/min with the 15.0 g mass on the plunger. Patients with glaucoma often have an outflow of less than 1 ml/min.

The Goldmann aplanation tonometer, developed about 1955, gives a more accurate measure of the ocular pressure. The measurement is usually taken with the patient in a sitting position (Fig. 12.39). The principle is simple; the force needed to flatten an area 3.06 mm in diameter on the front of the cornea is measured. The operator looks through the optical system and adjusts the small force needed to cause the desired flattening. The force needed for a normal eye is equivalent to the weight of a 1.7 g mass. This small force raises the internal pressure by about 65 Pa, while the Schiotz tonometer increases it about 2000 Pa. The Goldmann tonometer is calibrated directly in millimeters of mercury of internal pressure. The rigidity of the eyeball has little effect on the reading.

The Schiotz tonometer is seldom used, but the principle is used in a small pen-like instrument with a digital readout. This instrument, although not as accurate as the Goldmann tonometer, can be used on eyes which have had surgery of the cornea.

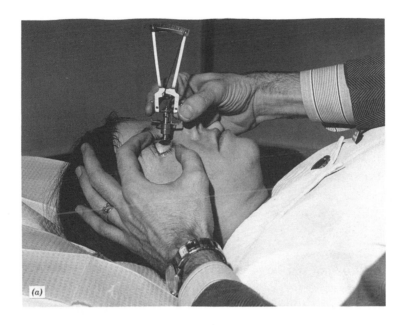

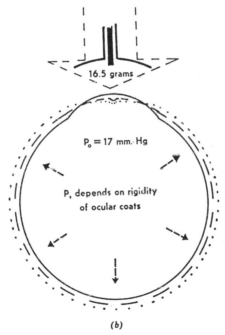

Figure 12.36. (a) The Schiotz tonometer measures the pressure in the eye by determining the deflection of the cornea under a given force, usually 16.5 g. (Courtesy of Thomas Stevens, M.D.) (b) Schematic of the Schiotz tonometer in use. The internal pressure before application of the tonometer is P_0. The pressure after the application of the tonometer P_t depends on the rigidity of the eye. (From Robert C. Drews, *Manual of Tonomgraphy*, C. V. Mosby Company, St. Louis, 1971, p. 10.)

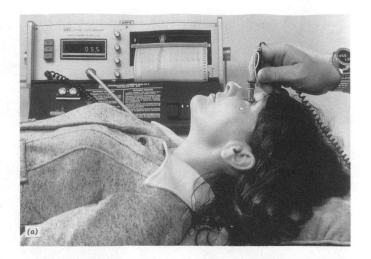

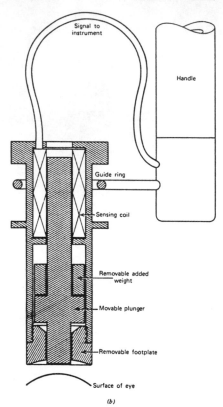

Figure 12.37. An electronic Schiotz tonometer. (a) The operator lets the measuring element rest on the subject's eye. The operator can read the pressure and the elapsed time on his or her wrist meter. (b) The weighted movable plunger that rests on the eye affects the magnetic field of the sensing coil to produce an electrical signal that indicates the intraocular pressure.

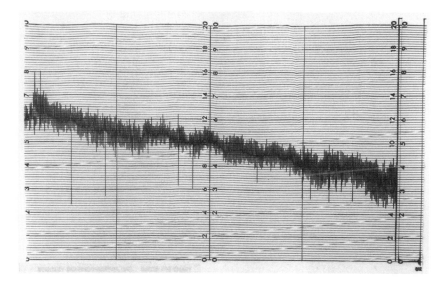

Figure 12.38. Tonograph showing the decrease in pressure with time. The slope permits an estimate of the flow rate. (Courtesy of Thomas Stevens, M.D.)

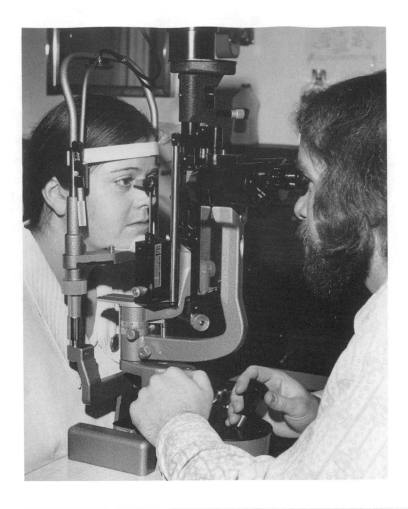

Figure 12.39. The Goldmann aplanation tonometer measures the force needed to flatten an area 3.06 mm in diameter on the front of the cornea. (Courtesy of Thomas Stevens, M.D.)

General and Chapter References

The following list of references is included as a guide to material which both supports and supplements the text. General references of interest to individuals using this text or which relate to many of the chapters in this book are listed first. References specific to individual chapters (2 through 12) follow.

General References

Baker, D. J., Jr., Resource Letter PB-1: Physics and Biology, *Amer. J. Phys.* **34** (2): 83–93 (1966).

Berg, H. C., *Random Walks in Biology* (Princeton Univ. Press, Princeton, NJ, 1983).

Bernstein, H. J., and V. F. Weisskopf, "About liquids," *Amer. J. Phys.* **55** (11): 974–983 (1987).

Connolly, W. C., "The physics of sport activities," *Phys. Teach.* **16** (6): 392–396 (1978).

Freeman, W. J., "The physiology of perception," *Sci. Amer.* **2**: 78 (1991).

Frolich, C., Resource Letter PS-1: Physics of Sports, *Amer. J. Phys.* **54** (7): 590–593 (1986).

Glass, L., and M. C. Macky, *From Clocks to Chaos: the Rhythms of Life* (Princeton Univ. Press, Princeton, NJ, 1988).

Hafemeister, D., Resource Letter BELFEF-1: Biological Effects of Low-Frequency EM Fields, *Amer. J. Phys.* **64** (8): 974–981 (1996).

Hobbie, R. K., Resource Letter MP-1: Medical Physics, *Amer. J. Phys.* **53** (9): 822–829 (1985).

Hobbie, R. K., *Intermediate Physics for Medicine and Biology*, 3rd edition (Springer-Verlag, New York, 1997).

Hobbie, R. K., "Teaching exponential growth and decay: examples from medicine," *Amer. J. Phys.* **41** (3): 389 (1973).

Hoop, B., Resource Letter PPPP-1: Physical principles of physiological phenomena, *Amer. J. Phys.* **55** (3): 204–210 (1987).

Lin, H., "Newtonian mechanics and the human body: some estimates of performance," *Amer. J. Phys.* **46** (1): 15–18 (1978).

Lin, H., "Fundamentals of zoological scaling," *Amer. J. Phys.* **50** (1): 72–81 (1982).

McMahon, T. A., and J. T. Bonner, *On Size and Life* (Scientific American Library, 1983) [...considers the implications of size and shape for organisms, beginning with a discussion of the role of size in natural selection...].

Patton, H. D., et al., (Eds.), *Textbook of Physiology, Vol. 1 and 2* (W. B. Saunders Company, Philadelphia, 1989). Vol. 1: Vol. 2: *Circulation, Respiration, Body Fluids, Metabolism, and Endocrinology.*

Purcell, E. M., "Life at low reynolds numbers," *Amer. J. Phys.* **45** (1): 3–11 (1977).

Ruch, T. C., and H. D. Patton (Eds.), *Physiology and Biophysics*, 19th edition (W. B. Saunders Company, Philadelphia, 1965). [See also Patton, H. D., et al. *Textbook of Physiology.*]

Stibitz, G. R., *Mathematics in Medicine and the Life Sciences* (Year Book, Chicago, 1966).

Tustin, A. "Feedback," *Sci. Amer.* **9**: 48–55 (1952).

Winfree, A. T., *The Timing of Biological Clocks* (Scientific American Library, 1987).

Chapter 2—Energy, Heat, Work, and Power of the Body

Bartlett, A. A., "The physics of heat stroke," *Amer. J. Phys.* **51** (2): 127–132 (1983).

Benzinger, T. H., "The human thermostat," *Sci. Amer.* **204**: 134–137 (1961).

Editors, *Food Intake and Energy Expenditure* (CRC Press, Boca Raton, 1994).

Johnson, A. T., *Biomechanics and Exercise Physiology* (John Wiley & Sons, Inc., New York, 1991).

Kearney, J. T., "Training the olympic athlete," *Sci. Amer.* **6**: 53 (1996).

Kleiber, M., *The Fire of Life: An Introduction to Animal Energetics* (John Wiley & Sons, Inc., New York, 1961).

Martin, R. J., B. D. White, and M. G. Hulsey, "The regulation of body weight," *Amer. Sci.* **79** (6): 528 (1991).

Ruch, T. C., and H. D. Patton (Eds.) *Physiology and Biophysics*, 19th edition (W. B. Saunders Company, Philadelphia, 1965). [See also Patton, H. D. et al., *Textbook of Physiology.*]

Saho, S. B., "Handling hypothermia," *The Science Teacher*, p. 25 (December, 1996). [Investigative experiments for the high school classroom.]

Webb, P., "Work, Heat, and Oxygen Costs," pp. 847–879 in *Bioastronautics Data Book*, J. F. Parker and V. R. West (Eds.) (NASA, Washington, DC, 1973).

Weinstock, H., "Thermodynamics of cooling a (live) human body," *Amer. J. Phys.* **48** (5): 339–341 (1980).

Whitt, F. R., *Bicycling Science: Ergonomics and Mechanics*, edited and enlarged by D. G. Wilson (MIT Press, Cambridge, MA, 1975).

Chapter 3—Muscle and Forces

Bellemans, A., "Drag force exerted by water on the human body," *Amer. J. Phys.* **49** (4): 367–368 (1981).

Cross, R., "Standing, walking, running, and jumping on a force plate," *Amer. J. Phys.* **67** (4): 304–309 (April, 1999).

Fung, Y. C., N. Perrone, and M. Anliker (Eds.), *Biomechanics: Its Foundations and Objectives*, Symposium on the Foundations and Objectives of Biomechanics, La Jolla, Calif., 1970 (Prentice-Hall, Englewood Cliffs, NJ, 1972).

Fung, Y. C., *Biomechanics: Mechanical Properties of Living Tissue* (Springer-Verlag, New York, 1981).

Fung, Y. C., *Biomechanics: Motion, Flow, Stress and Growth* (Springer-Verlag, New York, 1990).

Garfield, S., *Teeth, Teeth, Teeth* (Simon & Schuster, New York, 1969).

Herschman, A., "Animal locomotion as evidence for the universal constancy of muscle tension," *Am. J. Phys.*, **42** (9): 778–779, (1974). [See also Schmidt-Neilsen, K., *Science* **177**: 232 (1972). Gold, A., *Science* **181**: 275, (1973).]

Johnson, A. T., *Biomechanics and Exercise Physiology* (John Wiley & Sons, Inc., New York, 1991).

Laws, K., "Physics and dance," *Amer. Sci.* **73** (5): 426 (1985). [Also in *Physics Today.*]

Lenihan, John, *How the Body Works* (Medical Physics Publishing, Madison, WI, 1995).

Nordin, M., and V. H. Frankel, *Basic Biomechanics of the Musculoskeletal System*, 2nd edition (Lea & Febiger, Philadelphia, 1989).

Pedley, T. J. (Ed.), *Scale Effects in Animal Locomotion* (Academic Press, 1977).

Pritchard, W. G., and J. K. Pritchard, "Mathematical models of running," *Amer. Sci.* **82** (6): 546 (1994).

Stossel, T. P., "The machinery of cell crawling," *Sci. Amer.* **9**: 54 (1994).

Walker, J. D., "Karate strikes," *Amer. J. Phys.* **43** (10): 845–849 (1975).

Wilk, S. R., R. E. McNair, and M. S. Feld, "The physics of karate," *Amer. J. Phys.* **51** (9): 783–790 (1983).

Williams, M., and H. R. Lissner, *Biomechanics of Human Motion* (W. B. Saunders Company, Philadelphia, 1962).

Yamada, H., *Strength of Biological Materials* (edited by F. G. Evans) (Williams and Wilkins, Baltimore, 1970).

Chapter 4—Physics of the Skeleton

Fardon, D. F., *Osteoporosis: Your Head Start on the Prevention and Treatment of Brittle Bones* (MacMillan, New York, 1985).

Fung, Y. C., N. Perrone, and M. Anliker (Eds.), *Biomechanics: Its Foundations and Objectives*, Symposium on the Foundations and Objectives of Biomechanics, La Jolla, Calif., 1970 (Prentice-Hall, Englewood Cliffs, NJ, 1972).

Institute of Physical Sciences in Medicine Report 4 *Osteoporosis & Bone Mineral Measurement* (IPSM, York, England, 1988).

Kraus, H., "On the Mechanical Properties and Behavior of Human Compact Bone" in S. N. Levine (Ed.), *Advances in Biomedical Engineering and Medical Physics* (Wiley-Interscience, New York, 1968), Vol. 2, pp. 169–204.

McIlwain, H. H., *Osteoporosis: Prevention, Management, Treatment* (John Wiley & Sons, New York, 1988).

Nordin, M., and V. H. Frankel, *Basic Biomechanics of the Musculoskeletal System*, 2nd edition (Lea & Febiger, Philadelphia, 1989).

Peck, W. A., *Osteoporosis: The Silent Thief* (American Association of Retired Persons (AARP), Washington, DC, 1988).

Wright, V. (Ed.), *Lubrication and Wear in Joints*, Proceedings of a Symposium organized by the Biological Engineering Society and held at The General Infirmary, Leeds, on April 17, 1969 (Lippincott, Philadelphia, 1969).

Yamada, H., *Strength of Biological Materials*, edited by F. G. Evans (Williams and Wilkins, Baltimore, 1970).

Chapter 5—Pressure in the Body

Fundamentals of Hyperbaric Medicine, National Research Council Committee on Hyperbaric Oxygenation, Publication 1298 (National Academy of Sciences-National Research Council, Washington, DC, 1966).

New Science of Skin and Scuba Diving, 5th revised ed., Council for National Cooperation in Aquatics, New Century Publishers, Inc., Piscataway, NJ, 1980.

Chapter 6—Osmosis and the Kidneys

Hobbie, R. K., "Osmotic pressure in the physics course for students of the life sciences," *Amer. J. Phys.* **42** (3): 188–197 (1974).

Hobbie, R. K., *Intermediate Physics for Medicine and Biology*, 3rd edition (Springer-Verlag, New York, 1997).

Verniory, A., et al., "Measurement of the permeability of biological membranes," *J. Gen. Physiol.* **62**: 489–507 (1973).

Chapter 7—Physics of the Lungs and Breathing

Clements, J. A., *Surface Tension in the Lungs* (Freeman, San Francisco, 1962).

Haas, F., *The Chronic Bronchitis and Emphysema Handbook* (John Wiley & Sons, Inc., New York, 1990).

Moon, R. E., R. D. Vann, and P. B. Bennett, "The physiology of decompression illness," *Sci. Amer.* **8**: 70 (1995).

"Nature's Pumps," *Amer. Sci.*, **82** (5): 464 (1994).

Petty, T. L., *Enjoying Life with Emphysema* (Lea and Febiger, Philadelphia, 1984).

Phillips, M., "Breath tests in medicine," *Sci. Amer.* **267**: 74–79 (1992).

West, J. B., "Human physiology at extreme altitudes on mount everest," *Science* **223**: 784–788 (24 February 1984).

Chapter 8—Physics of the Cardiovascular System

Brown, B. A., *Hematology: Principles and Procedures*, 2nd edition (Lea and Febiger, Philadelphia, 1976).

Crane, H. R., "How to pump blood without a heart," *Phys. Teach.* **22** (4): 252–255 (1984).

Cromwell, L., F. J. Weibell, E. A. Pfeiffer, and L. B. Usselman, *Biomedical Instrumentation and Measurements*, (Prentice-Hall. Englewood Cliffs, NJ, 1973).

Debakey, M. E., *The Living Heart* (McKay, New York, 1977).

Ghista, D. N. (Ed.), *Advances in Cardiovascular Physics Vol. 6: Non-invasive Cardiac Assessment Technology* (Karger, New York, 1989).

Glass, L., P. Hunter, and A. McCulloch (Eds.), *Theory of Heart: Biomechanics, Biophysics, and Nonlinear Dynamics of Cardiac Function* (Springer-Verlag, New York, 1991).

Hobbie, R. K., *Intermediate Physics for Medicine and Biology*, 3rd edition (Springer-Verlag, New York, 1997).

Hollenberg, N. K., *The Heart Facts: What you can do to keep a healthy heart* [American Association of Retired Persons (AARP), Washington, DC, 1989].

Kenner, T., et al., *Cardiovascular System Dynamics: Models & Measurements* (Plenum, New York, 1982).

Nadel, E. R., "Physiological Adaptations to Aerobic Training," *Amer. Sci.* **73** (4): 334 (1985).

Chapter 9—Electrical Signals from the Body

Axel, R., "The molecular logic of smell," *Sci. Amer.* **10**: 54 (1995).

Caianiello, E., *Physics of Cognitive Processes* (World Scientific, New York, 1987).

Cohen, D., "Magnetic fields around the torso: production by electrical activity of the human heart," *Science* **156**: 652–654 (1967).

Cohen, D., J. C. Norman, F. Molokhia, and W. Hood, Jr., "Magnetocardiography of direct currents: S-T segment and baseline shifts during experimental myocardial infarction," *Science* **172**: 1329–1333 (1971).

Cromwell, L., F. J. Weibell, E. A. Pfeiffer, and L. B. Usselman, *Biomedical Instrumentation and Measurements* (Prentice-Hall, Englewood Cliffs, NJ, 1973).

Hobbie, R. K., *Intermediate Physics for Medicine and Biology*, 3rd edition (Springer-Verlag, New York, 1997).

Hobbie, R. K., "Nerve conduction in the pre-medical physics course," *Amer. J. Phys.* **41** (10): 1176–1183 (1973).

Hobbie, R. K., "Improved explanation of the electrocardiogram," *Amer. J. Phys.* **52**: 704–705 (August, 1984).

Geddes, L. A., *Electrodes and Measurement of Bioelectric Events* (Wiley-Interscience, New York, 1972).

Goodgold, J., and A. Eherstein, *Electrodiagnosis of Neuromuscular Diseases* (Williams and Wilkins, Baltimore, 1972).

Hämäläinen, M., R. Hari, R. Ilmoniemi, J. Knuutila, and O. Lounasmaa. "Magnetoencephalography—theory, instrumentation, and applications to noninvasive studies of the working human brain," *Reviews of Modern Physics* **65**(2): 418 (April 1993).

Katz, B., *Nerve, Muscle, and Synapse* (McGraw-Hill, New York, 1966).

Malmivuo, J., and R. Plonsey, Bioelectromagnetism: Principles and Applications of Bioelectric and Biomagnetic Fields (Oxford University Press, 1995).

Neumann, J. W., *Listening to Your Own Body: A Guide to Neurological Problems that Afflict Us as We Grow Older* (Adler & Adler, Bethesda, Maryland, 1987).

Chapter 10—Sound and Speech

Denes, P. B., and E. N. Pinson, *The Speech Chain: The Physics and Biology of Spoken Language* (Doubleday Anchor, Garden City, NY, 1963).

Deutsch, D., "Paradoxes of musical pitch," *Sci. Amer.* **8**: 88 (1992).

Flanagan, J. L., *Speech Analysis, Synthesis and Perception*, 2nd edition (Springer-Verlag, New York, 1972).

Fry, D. B., *The Physics of Speech* (Cambridge University Press, Cambridge, 1979).

Rigden, J. S., *Physics and the Sound of Music* (John Wiley & Sons, New York, 1977).

Rossing, T. D., *The Science of Sound* (Addison-Wesley Publishing Company, 1982).

Sataloff, R. T., "The human voice," *Sci. Amer.* **12**: 108 (1992).

Chapter 11—Physics of the Ear and Hearing

Allen, J. B., and S. T. Neely, "Micromechanical models of the cochlea," *Physics Today* **45**: 40–47 (1992).

Bess, Fred H., and L. E. Humes, *Audiology: The Fundamentals* (Williams and Wilkins, Baltimore, 1990).

Feldman, A. S., and L. A. Wilber, *Acoustic Impedance and Admittance: The Measurement of Middle Ear Function* (Williams and Wilkins, Baltimore, 1976).

Hudspeth, A. J., "How the ear's works work," *Nature* **341**: 397 (Oct. 5, 1989).

Hudspeth, A. J., "A cellular basis of hearing: the biophysics of hair cells," *Science* **230**: 745 (1985).

Hudspeth, A. J., and Markin, V. S., "The ear's gears: mechanico-electrical transduction by hair cells," *Physics Today* **47** (2): 22–28 (1994).

Hults, M. G., "Binaural hearing," *Phys. Teach.* **18** (7): 509 (1980).

Konoshi, M., "Listening with two ears," *Sci. Amer.* **4**: 66 (1993).

Milder, F., "Apparatus for demonstrating resonances of the inner ear," *Phys. Teach.* **16** (7): 503–504 (1978).

Peterson, J. E., "Hearing by bone conduction," *Phys. Teach.* **17** (8): 537–538 (1979).

Roederer, Juan G., *The Physics and Psychophysics of Music, An Introduction*, 3rd edition (Springer-Verlag, New York, 1995).

Rowell, N. P., "The response of the ear," *Phys. Teach.* **18** (7): 532 (1980).

Von Bekesy, G., *Experiments in Hearing* (McGraw Hill, New York, 1960). [The classic investigations on this subject.]

Chapter 12—Physics of the Eyes and Vision

Cornsweet, T. N., *Visual Perception* (Academic, New York, 1970).

Edge, R. D., "The optics of the eye lens," *Phys. Teach.* **27** (5): 392–393 (1989).

Ficken, Jr., G. W., "Estimating logarithmic base of eye's intensity response," *Phys. Teach.* **17** (2): 111 (1975).

Goldberg, F., Sharon Bendall, and Igal Galili, "Lenses, pinholes, screens, and the eye," *Phys. Teach.* **29** (4): 221–224 (1991).

Grant, R. M., "The red-eye effect," *Phys. Teach.* **23** (8): 514 (1985).

Kaiser, P. K., and R. M. Boyton, *Human Color Vision* (Optical Society of America, 1996).

Keating, M. P., "Reading through pinholes: A closer look," *Amer. J. Phys.* **47** (10): 889–891 (1979).

Legrand, Y., and S. G. El Hage, *Optical Sciences Series Vol. 13: Physiological Optics* (Springer-Verlag, New York, 1980).

Lock, J. A., "Fresnel diffraction effects in misfocused vision," *Amer. J. Phys.* **55** (3): 265–269 (1987).

Pease, P. L., Resource letter CCV-1: Color and color vision, *Amer. J. Phys.* **48** (11): 907–917 (1980).

Rock, I., *Perception*, (Scientific American Library, 1984).

Roorda, A., and D. R. Williams, "New directions in imaging the retina," *Optics&Photonics News*, 23 (February, 1997).

Rossotti, H., *Colour: Why the World Isn't Grey* (Princeton University Press, Princeton, NJ, 1988).

Sherman, P. D., and A. Kropf, "Color vision in the nineteenth century: the Young-Helmholtz-Maxwell theory," *Amer. J. Phys.* **51** (7): 670–671 (1983).

White, H. W., et al., "Undergraduate laboratory experiment to measure the threshold of vision," *Amer. J. Phys.* **50** (5): 448–450 (1982).

Zeki, S., "Visual image in mind and brain," *Sci. Amer.* **9**: 68 (1992).

Appendix A

Standard Man

In medical physics, where we are concerned with the anatomy and physiology of humans, it is convenient to define the physical characteristics of a "standard man." While the standard man is nonexistent, the following somewhat arbitrary values are useful for simulation and for computational purposes:

Age	30 yr
Height	1.72 m (5 ft 8 in.)
Mass	70 kg
Weight	690 N (154 lb)
Surface area	1.85 m^2
Body core temperature	37.0 C
Body skin temperature	34.0 C
Heat capacity	3.6 kJ/kg C (0.86 kcal/kg C)
Basal metabolism	44 W/m^2 (38 kcal/m^2 hr, 70 kcal/hr, 1680 kcal/day)
Heart Rate	70 beats/min

Blood volume   5.2 liters
Cardiac output   5 liters/min
Blood pressure—systolic   16 kPa (120 mm Hg)
Blood pressure—diastolic   10.5 kPa (80 mm Hg)
Breathing rate   15/min
O_2 consumption   0.26 liter/min
CO_2 production   0.21 liter/min
Total lung capacity   6 liters
Tidal volume   4.8 liters
Lung dead space   0.15 liters

	Mass (kg)	**% of body mass**
Muscle	30	43
Fat	10	14
Bone	7.0	10
Blood	5.4	7.7
Liver	1.7	2.4
Brain	1.5	2.1
Both lungs	1.0	1.4
Heart	0.3	0.43
Each kidney	0.15	0.21
Thyroid	0.02	0.029
Each eye	0.015	0.021

Appendix B

Exponential Behavior and Logarithms

The description of scientific phenomena can be simplified through the use of mathematics. One aspect of mathematics which is particularly important in medical physics involves an understanding of the behavior both of exponentials and of logarithms.

B.1 Exponential Behavior

Exponential behavior occurs when the rate of change of a quantity is proportional to the quantity initially present. Here is a short list of phenomena of interest in medical physics which behave this way:

1. Radioactive substances decay exponentially with time.
2. Growth of bacteria increases exponentially with time.
3. X-rays, gamma rays, and ultrasound each are absorbed exponentially with distance in the body.
4. The kidneys remove many substances exponentially from the body with time.

Let's use the growth of bacteria as an example. Suppose that it is found that a culture containing these bacteria increases in mass by 10 percent each hour. This increase in mass is the result of the division of individual bacteria into two identical bacteria (they bifurcate). If we call the initial number of bacteria B_0, then the number at the end of the first hour will be

$$B_1 = B_0 (1 + 0.10) = B_0 (1.10).$$

During the second hour, the amount will again increase by 10 percent, this time over the starting value of B_1:

$$B_2 = B_1 (1.10) = B_0 (1.10)(1.10) = B_0 (1.10)^2.$$

It follows that after h hours have elapsed, the quantity will be $B_h = B_0 (1.10)^h$.

[Note that the meaning of "exponential change" comes through here: to find the quantity after a certain time has elapsed, we raise the factor expressing the amount of change to be expected in one time period (the quantity grows by 10 percent in our example, or is a factor of 1.10 larger than at the start) by the number of elapsed time periods, h—the growth factor is exponentiated by h.]

To make this more general, let r represent the **rate** of growth rate in each fixed period of time, and y_0 be the initial quantity. The quantity present after N periods of time, y_N will be given by:

$$y_N = y_0 (1 + r)^N$$

Since all changes of interest in medical physics occur continuously with time (or with distance), this result must be generalized. Standard textbooks of calculus show that, for continuous growth at a rate, r, after time, t, has elapsed,

$$y_t = y_0 \, e^{rt}$$

where the base $e = 2.718$, the base of natural logarithms.

A similar analysis for cases where the quantity in question is **decreasing** at a constant rate, **r**, with time, **t**, or at a constant rate, **k**, with distance, **x**, shows that:

after time, t: $\qquad\qquad\qquad y_t = y_0 \, e^{-rt}$

after distance, x: $\qquad\qquad y_x = y_0 \, e^{-kx}$

A graph of the equation for y_x, for an initial $y_0 = 100$ and **k** $= 1$ is shown in Fig. B.1a plotted on conventional linear graph paper. The same graph is shown plotted on semilog graph paper in Fig. B.1b. The straight line on the semilog graph is characteristic of an exponential function (see below for an analysis of logarithms).

Summary of Properties of Exponentials:

For a particular base, a and exponents m, n:

(1) $a^m \cdot a^n = a^{(m+n)}$

(2) $a^m / a^n = a^{(m-n)}$

(3) $(a^m)^n = a^{m \cdot n}$

(4) $1/(a^m) = a^{-m}$

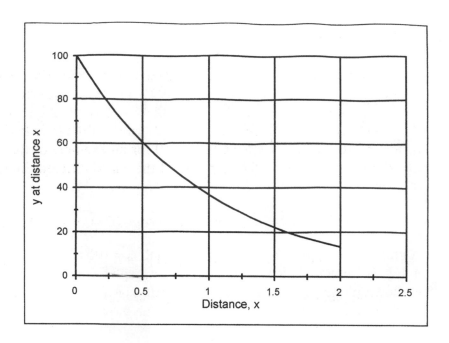

Figure B.1a. Linear plot, exponential decrease by distance.

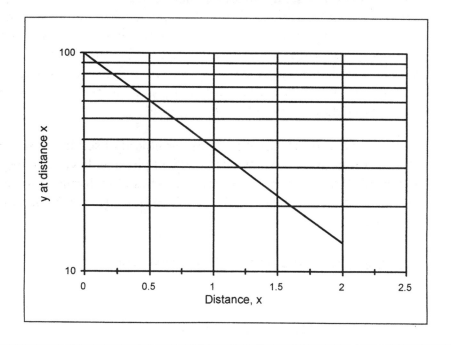

Figure B.1b. Semilog plot, exponential decrease by distance.

B.2 Logarithms

Logarithms are the inverse of the exponentiation process. The logarithm is important in medical physics because, for example, our sensory systems (e.g., our ears and our eyes) respond logarithmically to signal intensities which can range over 10 to 12 orders of magnitude (i.e., by factors of 10^{10} to 10^{12}). (See Chapter 10 for an example of the use of logarithms in describing the response of the ear to different intensities of sound.)

By definition, if $y = ax$, then x is the logarithm of y to the base a:

$$x = \log_a (y)$$

As an example, suppose the base, $a = 10$. Since $100 = 10^2$, then $2 = \log_{10}$ (100). In the same way,

$$3 = \log_{10} (1{,}000) = \log_{10} (10^3); \quad 4 = \log_{10} (10^4); \quad 10 = \log_{10} (10^{10}), \text{ etc.}$$

The semilog plot of Fig. B.1b is the logarithm, or mathematical inverse, form of the exponential equation we found in part B.1. If we take the logarithm of both sides of the equation for exponential decrease with distance, $y_x = y_0\, e^{-kx}$, we get $\ln (y_x/y_0) = -kx$, where ln denotes the natural logarithm to the base $\mathbf{e}$. The semilog plot is useful not only because it is linear for exponential behavior (note that x is proportional the number $\{\ln (y_x/y_0)\}$), but also because it can be used to plot quantities that cover several orders of magnitude.

Either base **10** or base **e** are most commonly used in cases of exponential or logarithmic behavior in medical physics. If you see **log y**, the base is **10**. If you see **ln y**, the base is **e**. These two forms of logarithms are related by the expression: $\ln y = (2.303) \log y$, which follows from the fact that $\ln (10) = (2.303)$.

Summary of Properties of Logarithms:

(1) $y = ax$; $x = \log_a (y)$

(2) $w = yz$; $\log_a (w) = \log_a (yz) = \log_a (y) + \log_a (z)$

(3) $w = y/z$; $\log_a (w) = \log_a (y/z) = \log_a (y) - \log_a (z)$

(4) $\log_a (yn) = n \log_a (y)$

Index